D0876478

The Forest and the City

Cecil C. Konijnendijk

The Forest and the City

The Cultural Landscape of Urban Woodland

Springer

Dr. Cecil C. Konijnendijk
woodSCAPE consult
Denmark

ISBN: 978-1-4020-8370-9 e-ISBN: 978-1-4020-8371-6

Library of Congress Control Number: 2008925102

Printed on acid-free paper

9 8 7 6 5 4 3 2 1

springer.com

For Mariska
For the forests

A tree on your doorstep, a forest in your mind.

Preface

During my childhood, nature was never far away. The vast polder landscapes near my home village had a lot to offer in terms of wildlife, space and adventure. One of my favourite places, however, was an abandoned fruit orchard which was conveniently situated next to our preferred football pitch. Here, my friends and I could pretend to be in the middle of a forest, playing hide and seek and, later, smoking a forbidden cigarette. Most importantly, the orchard gave us a feeling of just being away from everything. It offered an escape from homework, parental advice and even from the occasional broken heart.

As we lived in one of the least forested parts of The Netherlands, the 'real' forest was something exotic. It was the destination of our summer and autumn holidays, when my parents loaded up the car and drove us to the 'wild woods' of the Veluwe, the Belgian Ardennes, or perhaps even the German Black Forest. Forests were synonymous with freedom for me. They also meant watching wildlife in the dusk and the sights and sounds of a summer campground.

It was not before I started studying forest science that I realised that some people have the forest right at their very doorstep. People living in cities like Arnhem or Breda did not have to drive far to listen to woodpeckers and observe deer. Even nearby Rotterdam had its Kralingse Bos.

These city forests started to fascinate me. I had never really associated cities with forests, but soon realised that city and forest have gone hand in hand. This led to an academic interest in city forestry and to many travels and studies across the world. This book is part of my journey to understand city forests and to give them their rightful place in society. It takes a close look at our city forest heritage, with emphasis on the social and cultural dimensions of city forestry.

Although I have visited city forests in almost all parts of the world, this book is biased towards Northern and particularly North-western Europe, as these regions are the most familiar to me. I do believe, however, that many of the characteristics and roles of city forests that are described in this book are just as valid for a city forest near Tokyo as for the woodlands of Copenhagen. As nature on our doorsteps, city forests are pivotal in defining our relationships with forests and with nature in general.

This book derives from my professional as well as personal interest in city forests. It is based on many journeys, excursions, talks with city forest managers and users, as well as formal studies. It could never have been written without the contributions of the wider community of people interested in city forestry and urban green space. Very important in this respect has been my involvement with the European Forum on Urban Forestry, a network of city forest and urban green space planners, managers and scientists. My special thanks go to my colleagues in this network, as well as to all those people with whom I have communicated about city forestry.

Words of gratitude should also go to Catherine Cotton and Ria Kanters at my publisher Springer. Catherine was enthusiastic about the idea for this book from the very start, which was essential for making an idea into reality. Several colleagues kindly provided literature suggestions and illustrations. My thanks go to George Borgman, Janette Derksen, Hannelore Goossens, Rachel Kaplan, Sophia Mooldijk, Renate Spaeth, Jeroen Voskuilen, Moshe Shaler, and many others. Thanks also to Mersudin Avdibegović and Haris Piplaš for preparing information about the Vrelo Bosne area near Sarajevo.

I am much indebted to several good colleagues who were so kind to review one or two chapters. The suggestions of Marco Amati, Kevin Collins, Rik De Vreese, Robert Hostnik, Raffaele Lafortezza, Sylvie Nail, Anders Busse Nielsen, Janez Pirnat, Eva Ritter, Giovanni Sanesi, Alan Simson and Jasper Schipperijn were of invaluable help in improving the text. Having said this, the reviewers are in no way responsible for the final texts included in this book.

Some final words of thanks go to people that have provided essential support and inspiration before and during the writing process. Thanks, first of all, to my parents for taking me to the forests even before I could walk. Thanks also to my father-in-law, Hugo Goossens, for always showing an interest in my work and for a continuous flow of newspaper and magazine articles. Alan Simson, a dear friend and colleague, has been another great supporter throughout the writing process, always saying and writing the right things at the right time. Finally, I thank my dearest Mariska for always believing in me.

Dragoer, Denmark, *Cecil C. Konijnendijk*
December 2007

Contents

Preface vii

Brief Biography of Cecil C. Konijnendijk xiii

1 Introduction 1

Culture, Landscape and Forest 2
The Urban Era 4
Cities and Trees 5
Cities and Forests 7
City Forests as Cultural Landscapes, Place and Space 10
The City Forest Defined 13
Contents of This Book 15

2 The Spiritual Forest 19

Ancient Spiritual Links Between Forests and People 19
Forest and Religion 21
Forest and Myths 25
Modern Spirituality 29
Recreating Spiritual Links 32

3 The Forest of Fear 35

Primeval Forest Fear 35
Criminal Forests 38
Nature's Dangers 42
From Fear to Excitement 44

4 The Fruitful Forest 49

Subsistence Forests 49
Appropriation of City Forest Use 52
City Forests and Societal Development 54

The Rise of Municipal Forestry 56
The Fruitful Forest Today 60

5 The Forest of Power 63

From Wooded Commons to Elitist Prestige 64
Military Forests 68
City Forests and Democratisation 71
City Forests and City Image 74
City Forests and Environmental Justice 78

6 The Great Escape 81

Antidote to the City 81
Recreate and Enjoy 83
Window to the World 86
New Ways of Escaping 87
Different People, Different Escapes 90
Escape from City Rule 94

7 A Work of Art 97

City Forests as Inspiration 97
Designed Landscapes 100
Settings for Art 104
Art for Developing Community-Forest Links 106

8 The Wild Side of Town 111

City Forests and Nature Conservation 112
Nature Is in the Eye of the Beholder 117
Searching for the Urban Wilderness 121
Working with Nature 123

9 The Healthy Forest 127

Views on Nature and Health 128
City Forests and Physical Health 131
City Forests and Mental Health and Well-Being 135
Promoting the Healthy Forest 138

10 The Forest of Learning 143

City Forests as Testing Grounds 143
International Inspiration 147
Demonstrating Forestry to the Public 149
City Forests, Nature Education and Children 153
Tomorrow's Forest of Learning 157

11 The Social Forest ... 161

From Wooded Commons to Forests for All ... 161
Social Forests for Mass Use ... 165
City Forests as Social Stages ... 168
Social Forests and Place Making ... 170
Building Community Identity ... 172

12 A Forest of Conflict ... 177

Urban Development Conflicts ... 178
Conflicts over Forest Management ... 181
Recreation Conflicts ... 182
Other Conflicts ... 185
Managing the Forest of Conflict ... 186

13 A Forest for the Future ... 189

Challenges in an Urban Era ... 190
Developing Attractive Urban Landscapes ... 192
Building Communities ... 198
Maintaining the Links with Nature ... 202
Final Thoughts ... 205

Literature ... 207

Index ... 233

Brief Biography of Cecil C. Konijnendijk

Dr. Cecil Konijnendijk, a Dutch national based in Denmark, has studied and promoted the role of woodlands and trees in urban societies throughout his career. After employment with the European Forest Institute and the Danish Centre for Forest, Landscape and Planning, Cecil started woodSCAPE consult in 2004. His present work includes, among other, research, teaching, training of urban forestry professionals and writing about urban forestry issues. He has coordinated several international networks and research projects dealing with city forests and other green space. Cecil is also editor-in-chief of the scientific journal Urban Forestry & Urban Greening.

Chapter 1
Introduction

Djurgården and Stockholm, Jægersborg Dyrehave and Copenhagen, Epping Forest and London, Grunewald and Berlin, Bois de Boulogne and Paris, the Amsterdamse Bos and Amsterdam – these examples all show that over time, close relations between cities and forests have developed. 'City forest' is a term that exists in many different languages. Stadtwald (German), stadsbos (Dutch) and byskov or bynær skov (Danish) are only some examples of this. Traditionally, a city forest has been defined as a forest owned and/or managed by a certain city. Gradually the scope of the city forest concept has been broadened, now referring to a forest situated in or adjacent to a city, regardless of ownership. Perhaps more characteristic, however, are the close links between a forest and the local urban society (Konijnendijk, 1999). Today, city forests are often treated as part of a larger urban and peri-urban green structure. The term 'urban forest' has come into wider use, for example, to refer to the planning and management of all tree-dominated green resources, both publicly and privately owned, in and near an urban area (Randrup et al., 2005; Konijnendijk et al., 2006). Like trees along streets and in parks and gardens, city forests are one element of this wider 'urban forest'. In many cities, woodland plays a very important role, not in the least from a social and cultural perspective, within the overall urban forest and urban landscape.

This book looks at city forests primarily from a cultural and social perspective. They have developed over time in a close dialogue between the natural landscape and local, urban society. While city forests have been shaped by urban societies according to changing power relations and demands, they have in their turn impacted these societies, becoming part of local cultures and local identity. City forests also represent two important – contrasting but supplementary – concepts for analysing interactions between people and environments, namely those of place and space. Where place refers to home, familiarity and safety, space stands for the unknown, the wild, the adventurous (Tuan, 2007).

This introductory chapter sets the stage for the remainder of the book. First of all, it briefly looks at the links between culture, landscape and forest over time. Next, it describes the importance of urbanisation and the development of cities in changing the relationships between society, nature and forests. Links between cities, nature and green space are explored, after which the focus is directed towards city forests as cultural landscapes, as places and spaces. Finally, the contents of this book are introduced.

C.C. Konijnendijk, *The Forest and the City: The Cultural Landscape of Urban Woodland*

Culture, Landscape and Forest

Everything humans do, and our ideas of the natural world, exist in a context that is historically, geographically and culturally particular, and cannot be understood apart from that context. In order to understand a certain city society and its particular culture, it is thus important to understand its setting. The UNESCO Universal Declaration of Cultural Diversity (UNESCO, 2001) defines culture as a "set of distinctive spiritual, material, intellectual and emotional features of society or a social group, and that encompasses, in addition to art and literature, lifestyles, ways of living together, value systems, traditions and beliefs." In line with this, cultural values relate to what a society believes to be good, right and desirable.

Culture and nature are complex concepts, especially when related to one another. Culture is not only an opposing, dominating force with regards to nature. Where nature and culture meet, landscapes emerge. The European Landscape Convention (2000, p. 3) defines landscape as "an area, as perceived by people, whose character is the result of the action and interaction of natural and/or human factors". Landscapes are complex phenomena, as Schama (1995, p. 61) explains: "Landscapes are culture before they are nature; constructs of the imagination projected onto wood and water and rock. [...] But it should also be acknowledged that once a certain idea of landscape, a myth, a vision, establishes itself in an actual

Fig. 1.1 Landscapes are the result of complex interactions between nature and culture. Our preference for half-open landscapes such as this scenery from the Grizedale Forest in the English Lake District can be understood from analysing these complex interactions over time (Photo by the author)

place, it has a peculiar way of muddling categories, of making metaphors more real than their referents; of becoming, in fact, part of the scenery." Thus, Schama (1995, p. 14) also states that "our entire landscape tradition is the product of a shared culture, it is by the same token a tradition built from a rich deposit of myths, memories, and obsessions".

In order to understand cultures and societies, we need to have knowledge of the natural environment in which they have operated (Gifford, 2002), as well as of the landscapes that have resulted from the interactions between culture and the natural environment (Fig. 1.1). Our relationship to the natural environment has changed dramatically over time. The domestication of fire and later the shift to agriculture were important 'ecological transformations' in terms of man's relationship to nature, as we moved from being participants in nature to a position of power and dominance (Ponting, 1991; Goudsblom, 1992). Initial human domination over the natural environment was small-scale, but with the rise of civilisations and population growth, nature became overexploited. Forests, for example, gradually disappeared from the immediate surroundings of the capitals of former great civilisations, such as the Babylonian, Greek and Roman (Perlin, 1989). Later, during the times of Renaissance and Enlightenment, countryside and nature became subjects of a tendency of seeing nature not only from a functional point of view, but also as a source of aesthetics and recreation (e.g. Thomas, 1984).

In spite of societal moves of (re-)appreciation of nature, a third ecological transformation, i.e. the Industrial Revolution and its use of fossil fuels (Ponting, 1991), led to further exploitation of nature as well as to a growing distance between man and nature. The Industrial Era and its consequences for life in cities resulted in the elite expressing a renewed desire to 'go back to nature' (Hennebo, 1979; Van Rooijen, 1990).

This book looks at city forest landscapes. Like landscape, 'forest' is a complex concept as well, embodying natural as well as cultural and political dimensions. The German word 'Wald' is derived from wilderness and thus can be said to symbolise nature (e.g. Oustrup, 2007). Rackham (2004) and Muir (2005) use the term 'wildwood' when they speak of the unmanaged woodland. 'Forest', however, has a clear cultural component. Forest is derived from the Latin 'foris', i.e. "the area that lies outside the normal laws" (Rackham, 2004, p. 164). It was often used in the past as referring to unenclosed open land. However, the meaning of forest has changed over time, in line with societal changes and changing use of wooded lands. During the Middle Ages, forests started to refer to lands for which the King or Emperor held the exclusive hunting rights (Rackham, 2004; Muir, 2005). First forest laws concerned boundaries, pasturage, the right of feeding swine on mast, damages done to trees, and so forth. Walden (2002) describes how in the German language domain, 'Forst' – which initially referred to royal hunting areas – gradually came to refer to production forests, i.e. wooded areas primarily used for producing timber. 'Wood', which is derived from 'ved' and 'gwydd', commonly has referred to a collection of trees (Oustrup, 2007). 'Woodland' is used by Rackham (2004) to refer to all woods that are named, owned, given boundaries and most importantly managed, in contrast to 'wildwood'.

The Urban Era

Modern relationships between nature, forest, people and culture have mostly been shaped in our cities which testify "to our ability to reshape the natural environment in the most profound and lasting ways" (Kotkin, 2005, p. xviii). Or, in the words of Handlin (1963, p. 63): "In the modern city, the contest between the human will and nature assumed a special form". Many experts see urbanisation as the most important environmental influence on behaviour, as well as the most important behavioural influence on the environment (Gallagher, 1993). Cities shape people, whilst people shape cities.

'City' is defined in Merriam-Webster Online Dictionary (2007) as "an inhabited place of greater size, population, or importance than a town or village". Cities have always been important centres of civilisation and culture. Yet, current urban society is a fairly recent phenomenon. Only some ancient cities, such as Constantinople and Rome, hosted more than several thousands of inhabitants, while until very recently, the large majority of the world's inhabitants lived in rural areas. In 1800, for example, only one out of five people in Britain and Holland lived in cities; today about 90% of their population live in cities and towns. Berlin grew from 170,000 in 1800 to over 1.5 million in 1900 (Ponting, 1991). Urbanisation really took off during the Industrial Era, when cities attracted large influxes of workers for rapidly developing industries. Today, more than half of all people live in cities and towns. By the year 2030, 60% of the world's population is expected to be urban (UN DESA Population Division, 2005).

In describing the rise of the modern city, Handlin (1963) stresses the importance of the development of a centralised national state, a new productive system and vastly improved communications. Destruction of older – rural – ways of life was achieved through a convergence of political and economic forces. City life became dependent upon complex and impersonal arrangements of individuals. The way of life in modern cities, which uprooted traditional cultures, created grave social and personal problems. Anti-social behaviour and various social conflicts called for more governmental control. However, the rise of the modern city also led to the emergence of citizenship and a sense of civic consciousness.

With urbanisation, new patterns also emerged for disposing of space. Urban space does not just 'exist'; it is produced, reproduced and shaped in people's actions (Clark & Jauhiainen, 2006). Over time, city space came to be used in a much more specialised way, with functions being divided. Space – and also time – was reallocated with an eye towards its most profitable use. Space and time and society became intertwined. In line with this, Gehl (2007) mentions 'moving' (between spaces, functions) as one of the three key elements of cities over time, together with 'meeting' and 'market'.

Cities have been created in myriad forms over the past millennia (e.g. Handlin, 1963; Kostof, 1999; Kotkin, 2005). Some were developed from villages, while others have reflected the conscious vision of a high priest, ruler or business elite, following a general plan to fulfil some greater divine, political or economic purpose. In spite of cities being centres of art, religion, culture, commerce and technology, not all of them succeed. Kotkin (2005) lists three critical factors that make cities great – or, if not fulfilled, lead to cities' gradual demise. The sacredness of place, first of all, refers to religious structures that have long dominated the landscape and imagination of great cities. Cities were sacred places, connected directly to divine forces controlling the world. In our own, more secularly-oriented times, cities seek to create the sense of sacred place through towering commercial buildings and evocative cultural structures. Secondly, the need for security looks at cities as places of refuge from marauders and general lawlessness. The role of commerce, finally, relates to the production of wealth to sustain large populations for a large period of time. This requires an active economy of artisans, merchants and working people.

These ancient fundamentals are still important today. However, times have changed for the city. In an era of globalisation, decentralisation and new technologies that have led to diminishing advantages of scale, cities have become less important as economic centres. The ephemeral city has emerged, where the focus is on the role of cities as cultural and entertainment centres, for example for tourists (Kotkin, 2005; cf. Burgers, 2000). However, more is needed than constructing the "diversions of essentially nomadic populations" (Kotkin, 2005, p. 154). Successful cities also need a strong base in the form of a committed citizenry. This means that small business, schools and good neighbourhoods are just as much needed as museums, restaurants, theme parks, and so forth.

Driving forces in modern cities include that of power of representation, related to the image that the political powers want to create for themselves in cities, as well as those of commercial powers, resulting in office landscapes and trading/shopping streets. Migration is a third driving force, leading to an influx of people, talents, cultures and the establishment of new neighbourhoods. Tourism, as a fourth driving force, has a 'conserving' influence on cities, draining preserved city centres of normal city life and turning them into historical stages. Finally, the fifth driving force is fear, which has an enormous influence on how urban space is designed to optimise (feelings of) security (Ivfersen, 2007).

The urban ideal has demonstrated a remarkable resilience, being built up over and over again after wars and disaster. What really matters in the end is the peculiar and strong attachment that citizens have to a city and the sentiments that separate one specific place from another. The city is based on shared identity, on issues of community. It needs to inspire "the complex natures of gathered masses of people" (Kotkin, 2005, p. 160). 'Good' and successful cities still need to be sacred, safe and busy. They need to offer ample opportunities for dwellers and visitors to live, work, relax, to feel at ease and to be challenged and entertained. As this book will show, city forests have their role to play when the good city is at stake.

Cities and Trees

In line with the continuous production, reproduction and shaping of urban space in general, green space and its functions have also changed according to the preferences and demands of urban societies (e.g. Clark & Jauhiainen, 2006). From a societal view, the study of the development of urban green space helps illuminate the physical, socio-cultural and ecological changes in European cities, as well as the resulting ideas and policies regarding green and nature.

In the public landscape of cities, trees have been used in two main settings. First of all, there have been spaces used for public activities that the presence of trees enhanced, such as parade routes, areas for recreation, promenades for pedestrians and public spaces that function as squares or parks. Secondly, trees have been used as extensions of the private garden – and most often as street trees in front of houses. But all of these uses seem to have been almost entirely unknown before the middle of the 16th century in European cities (Lawrence, 2006). During the Middle Ages, trees primarily belonged to countryside, monastery gardens and some private gardens, the latter often in the form of walled gardens ('horti conclusi'). These private gardens, however, could be extensive and have the character of a 'semi-public park'. This was the case in European cities under Arabic influence, such as Palermo (Fariello, 1985).

Lawrence (2006) distinguishes three primary dimensions of the presence of (public) city trees over time: aesthetics, power and control, and national identity. The aesthetical dimension of trees, first of all, has been reflected in fashion, for example in designed gardens, green spaces as social venues, as well as in the choice of tree species. Over time, changes in aesthetics of architecture, garden design and urbanism influenced the ways trees were planted in cities. Changes in recreational activities also had an impact. Gradually, city trees became symbols of pleasure, fitting well into the emergence of the ephemeral city described earlier. Their presence made urban landscapes a more pleasant place to be, enhancing urban liveability. Trees played a crucial role in aristocratic leisure and later in a bourgeois sense of gentility, of a kind of refined and respectable pleasure. Aesthetics also has an ecological link, with trees as symbols of nature and linking to changes in the relationship between the human and natural worlds.

Secondly, trees have been deliberately planted for certain uses, and the ability to plant trees has reflected a degree of social power. Recreational spaces were not always freely open to all people. Shifting power relations between social classes and changing concepts of public and private realms have all influenced the planting and use of city trees. Government intervention increased steadily over time, catering for the leisure needs of the working classes. Until the 19th century, however, city trees mostly were a private matter. Private gardens became the first quasi public places; sometimes they were open to parts of the public. Those establishing green spaces (or previously, opening private parks and gardens) had a clear social agenda. Social use was based on a code of behaviour developed first by the aristocracy and then modified by the bourgeoisie – i.e. that of spaces of gentle, passive recreation.

Finally, from the perspective of national identity and fashions, the establishment and use of trees can be characterised as the gradual replacement of national styles of urban landscapes with a cosmopolitan set of urban landscape forms. Until the 19th century there were distinct ways of planting and using trees in cities in different parts of Europe (and America). Tree-lined canals, for example, were specific to The Netherlands (Fig. 1.2), formal-tree-lined boulevards to France, and enclosed residential squares to Britain. Later, these national peculiarities

Fig. 1.2 Trees planted along canals have been a distinct feature of Dutch cities such as Amsterdam (Photo by Moshe Shaler)

disappeared, as 'green space fashions' became Europe-wide due to increased travelling (primarily by the elite) and the boom in book and magazine publishing that offered inspiration.

These three key dimensions can be connected to the 'critical factors' for city success described in the previous section. For example, aesthetics is very much related to attractive, busy cities, while power has much to do with safety, as well as with economy and commercial interest. National (or city) tradition has a link to the sacred, to cities as places of community and identity.

By the mid-18th century, trees had become elements in urban improvements in response to the Enlightenment ideas of the human environment. Trees and green spaces became tied to programmes of reform which were focused on improving urban infrastructure, commerce, transportation and human health. Industrialisation in the 19th century and its associated challenges brought the most thorough changes in the use of trees in the urban landscape. The exploding population, poor living conditions of the working masses and a new social and political order led to the serious reform of urban infrastructure and green space.

Cities and Forests

Trees thus have played their role in city development, providing local identity and making important contributions to sacred, liveable, attractive and busy cities. How do city forests fit into all of this? Forests, of course, preceded most cities. Then, as cities developed, they were heavily dependent on surrounding woodland for the provision of construction wood, fuelwood, as well as fuel and raw materials for various industries (Perlin, 1989). This led to overexploitation and clearing of forests. The once densely forested surroundings of ancient cities like Knossos and Athens lost their trees while their civilisations (at least initially) boomed. Simon (2001) mentions a map of 1571 which shows the town of Belfast, Northern Ireland, still surrounded by extensive woods. All forests were subsequently cleared, until only few wooded areas remained.

The differences in the level and speed of urbanisation contributed to the emergence of regional 'forest cultures' in Europe (Bell et al., 2005). Urbanisation, particularly under the North-western and Southern forest culture, meant that surrounding areas were cleared of forests, unless these forests had a special protected status. Cities in the Nordic and Central forest culture, however, were more successful in conserving their forests. Nordic cities were often built 'into' the forest, thus creating a 'city in the forest'. Under the Central European forest culture, woodlands were appreciated for their many benefits and seen as an important part of national and local culture and identity. Thus many city administrations, for example, actively sought to protect nearby forests.

This book will show that, as in the case of city trees, aesthetics, power and national culture have all had their influence on city forests. Aesthetics are closely associated with changing recreational uses and preferences. The way in which city forests were transformed from more or less natural woods to hunting domains, and later, from hunting domains to recreational areas, reflects fashions of the time. The ecological role of city forests has been more important than that of most other city green space, as they have always represented nearby nature, a 'wilderness' at the urban fringe. The nature conservation movement emerged from a class of well-to-do, academics and artists living in cities, and who were often concerned about the nearby woods.

Power relationships have been reflected, for example, in rulers and the elite closing off city woods for their own (hunting and recreational) benefit. This conflicted with the traditional inclusion of many forests in the so-called (wooded) commons. Also, the elite were in the position to establish new woods, for example as part of their estates. Later, with the democratisation movement that swept across Europe from the mid-19th century onwards, most city forests came under municipal ownership, with the explicit aim to provide forest environments to all citizens.

National and local cultures also played their part. City forests raised the prestige of their owners, not in the least by acting as hunting domains. Like trees, they reflected the culture of their country, for example in their particular design. Over time, city forests were often seen as 'remnants' or representations of the natural landscape of a country, helping to build national identities.

Although city forests thus have undergone similar processes of change as city trees and urban green space at large, they have occupied a rather distinct place in the development of cities and their green structures. Part of this role relates to city forests representing the forest as the ancient base of civilisation, as 'the other' to the city, and as the wildest part of urban nature, offering a place of escape located at a convenient distance from people's homes.

The relation between cities, including their more 'urbane' forms of urban green space, and forests has been complex. On the one hand, the two have been at odds, in particular when urbanisation leads to the clearing of forest lands. On the other hand, as mentioned, forests have provided cities with essential products and services. The difficult relationship between the two is reflected in what Hibberd (1989, p. 18) writes in the case of Britain: "woodland has seldom been regarded as an essential part of the urban landscape". It is also reflected in the criticism of the design of a London square by landscape architect Humphry Repton: "In the due attention to the training and trimming of such trees by art, consists the difference between a garden and a park, or a forest; and no one will, I trust, contend that a public square should affect to imitate the latter." To the dislike of landscape architect Frederick Law Olmsted Jr. (Houle, 1987, p. xii), the forests around rapidly developing cities such as Portland, Oregon were regarded as "a 'troublesome' encumbrance standing in the way of a more profitable use of the land".

Yet, forests also provided inspiration, even to urban architects. For example, in his 'Essai sur l'architecture' (1753, cited in Hennebo, 1979, p. 77), Laugier stated that the city had to be viewed as a forest. City streets could be organised like forest roads and the star forests of hunting woods could be translated into city plans. The (designed) forest provided the organisation principle for the modern city. Also, in countries with cultures in which the forest continued to play a key role, such as Finland, forests had a strong influence on architecture and the making of cities (Treib, 2002). Treib cites the interview by a Danish journalist of the famous Finnish architect Alvar Aalto (in 1940) about city architecture and planning. In this interview Aalto said that nobody should be able to go home without passing through a forest.

Cities have sometimes been so green that they were mistaken for forests. In the 17th century, visitors to Dutch cities often wrote that they were not sure if they were seeing a city in a forest or a forest in a city. After his visit to Amsterdam in 1641, John Evelyn wrote: "Nothing more surpriz'd me than that stately, and indeede incomparable quarter of the Towne, calld the Keisers-Graft, or Emperors Streete, which appears to be a Citty in a Wood, through the goodly ranges of the stately and umbrageous Lime-trees, exactly planted before each-mans doore." (cited in Lawrence, 2006, p. 44).

Relatively few studies have been carried out on changing ideas, policies and attitudes related to urban green space (Clark & Jauhiainen, 2006). Trees and woods, for example, have been mostly regarded from a botanical and taxonomic perspective, even though they are important elements in the historical landscape (Muir, 2005). City forests in particular have often 'fallen in between' accounts of the development cities and urban green space on the one hand, and of forest history on the other (e.g. Walden, 2002). This may seem strange, given the importance of city forests in city development. Madas (1984) mentions several reasons why urban forests, including urban woodland, are such interesting phenomena to study. For example, small areas of urban woodland have to meet the demands of a very large number of inhabitants. Moreover, many cities have experienced a deteriorating quality of urban and suburban landscape in recent decades.

Earlier work has attempted to summarise some of the 'main lines' of city forestry history in Europe (Konijnendijk, 1997, 1999). Fortunately interesting studies and literature do offer information to fall back on, although most works are limited to a single city, a city forest or specific aspects of city forestry. In Germany, authors such as Brandl (1985, for Freiburg), Cornelius (1995, for Berlin), Walden (2002, for Hamburg) and Borgemeister (2005, for Hanover and Goettingen) have written excellent historical accounts of city forestry. The history of specific city forests has been published more frequently, as in the work of Derex (1997a, b) on the Bois de Boulogne and Bois de Vincennes of Paris, France. We do lack, however, comprehensive accounts of Europe's city forest heritage.

Although this book also has a strong historical component, its ambition is different. It aims to analyse city forests as cultural and social landscapes, focusing on the many essential links that cities and 'their' forests have had over time. Local society, local communities and their views towards the city forests are placed centrally. It is argued that the cultural and social links between local city dwellers and 'their' forests lie at the heart of the city forest concept, which has such as strong heritage in Europe (e.g. Hossmer, 1988). A better understanding of the cultural and social roots of the city forest concept will help us develop and manage city forests that cater for the demands of modern times, and help create better cities. Therefore, throughout the book, links are continuously made between the past, the present and the future of city forestry. Although cases from across Europe and from other parts of the world are described, this book has a bias towards North-western and Northern Europe, as these parts of the world are the most familiar to the author.

City Forests as Cultural Landscapes, Place and Space

In this book, city forests are studied as cultural landscapes. According to Arntzen (2002), a cultural landscape is seen as the result of an encounter between nature and culture, as some form of integration between the human and cultural with the natural, the land. Different types of experts take differing views of what a cultural landscape encompasses. Cultural and local historians, for example, focus on the visible aspect of a community's historical development, while cultural geographers look at aspects such as results of land use and tenure. In cultural landscapes, human contributions to the land can be constructive and consistent with nature's own conditions and processes, but this is not necessarily the case.

Arntzen (2002) makes a distinction between material cultural landscapes, which are physical and visible manifestations of the lives and activities of human communities, and immaterial cultural landscapes that are cultural in a spiritual or symbolic sense, on account of the significance they embody for members of a culture, even if they do not have visible, physical traces of human activity. "Cultural landscapes are the doings of human communities, and human communities have originally been attached to specific places or regions whose conditions have been central to the determination of the community's character and its impact on the land."

(Arntzen, 2002, p. 37). Not only is a cultural landscape influenced by humans, it also influences people's views of themselves, their present and past. Cultural landscapes hold a certain identity value as they somehow represent or embody people's identity at the local, regional or national level as members of an ethnic group. A cultural landscape has its own complex, distinct narrative, a story of its development.

Cultural forest landscapes, that is, landscapes with managed forests as the key element, have been crucial in the development of local and national societies. They have been essential sources of timber and fuelwood (Perlin, 1989), but have also contributed to the shaping of culture and national identity (Schama, 1995). Among Europe's cultural forest landscapes, forests near cities and towns have been especially important, as they were situated closest to cities which in themselves were driving forces of societal and cultural change. Thus these city forests represent a 'frontier area' for studying the (manifestations of) cultural links between society and forest.

City forest landscapes can also be considered from an experiential perspective, focusing on the concepts of space and place (based on Tuan, 2007). Space is a common symbol of freedom in the Western world. Space lies open, suggests the future and invites action. On the negative side of this, space also can hold a threat, as open and free can also mean exposed and vulnerable. In contrast to space, place can be characterised as enclosed and humanised space, as the calm centre of established values. In brief, place is security and home, and space is freedom and the unknown. Humans require both, as we are attached to the one (place) and long for the other (space), moving between shelter and venture, between attachment and freedom. We welcome both the boundedness of place and the exposure of space. Moreover, "when space feels thoroughly familiar to us, it has become place" (Tuan, 2007, p. 73).

Meanings assigned to a place are unique to that place and do not readily transfer to other places, even if the biophysical attributes are identical (Cheng et al., 2003). 'Sense of place' refers to profound psychological and emotional links that develop between people and the place they live in and perceive, experience and value (Cheng et al., 2003; Tuan, 2007). Sense of place relates to the biophysical context of the natural and built environments, as well as to knowledge, values and attitudes towards and about the surrounding environment. Psychological elements of sense of place refer to an individual's psychological connection to and physical dependence on place, dealing, for example, with place identity. Socio-cultural elements set sense of place in a social and cultural context, dealing with the influence of social community and cultural context on people-place relationships. Political-economic elements of sense of place, finally, reflect localised processes that arise in response to influences of the particular place and socio-cultural milieu (Ardoin, 2004).

The socio-cultural elements or dimension of sense of place is the focus of this book, although the other elements will also feature prominently. The socio-cultural dimension of place has to do with the 'inscription' of sense of place through cultural processes, social networks within place, as well as political and environmental involvement (activism) (Fig. 1.3). The cultural dimension refers to the exploration

Fig. 1.3 Sense of place refers to close psychological and emotional links between people and their environment. 'Place' is often rooted in social interaction – people jointly define and shape their favourite places. In this process, trees are often important 'place enhancers', as in the case of this Italian city square (Photo by the author)

and recognition of symbols that social groups use to produce and reproduce narratives about their places, such as city forests.

Social scientists have argued that connections with places are often strongest and most compelling when rooted in social interactions within a place, with the sense of a place deriving from people, experiences and memories created in the place (Ardoin, 2005). These ties relate to what the field of environmental psychology describes as place identity (Proshansky et al., 1983) at a more individual level, and as definitions of social or primary territory where groups are concerned (e.g. Gifford, 2002). Place identity refers to a strong sense of attachment to place. Most people retain a strong sense of attachment and belonging to key places. Definition of 'the self' might reflect relations with places as much as it reflects relations with people (Walmsley & Lewis, 1993). Territoriality operates at a community level to encourage group identity and bonding. In this context, Irwin Altman (cf. Gifford, 2002) defines primary territories as areas 'owned' by individuals or primary groups. These are controlled on a relatively permanent basis by them, and are central to their daily lives. Public territories, on the other hand, are open to anyone in good standing with the community. Occasionally, through discrimination or unacceptable behaviour, public territories are closed to individuals.

How do trees and forests relate to space, place, place identity, territoriality and related concepts? Forests, Tuan (2007) writes, can be associated with a sense of spaciousness. Although the cluttered forest environment may seem the antithesis of open space, it is in fact a "trackless region of possibility" (Tuan, 2007, p. 56). The forest as wilderness pre-dating civilisation and city is 'space'. With the clearing of forests and the development of the man-made world, however, the landscape becomes well defined and more 'placey'.

But it would be wrong to see forests only as space, as wild and unknown areas of venture, exposure and freedom. Especially in the case of city forests, the woods also can be places, being part of local people's daily living environment. Treib (2002) stresses the importance of trees in place making, for example by helping to physically define space and suggest a specific location. The role of trees in place making is also described by Jones and Cloke (2002). They see the concept of place as significant in the construction of the everyday world and as a dynamic and shifting phenomenon. Places are, in part, dynamic outcomes of the coming together of local, regional, national and international cultural constructions, in co-present material, social, economic and historical contexts (Jones & Cloke, 2002). The authors state that place, however, is not the same as (human) community, as non-human actors such as trees, play an important role as 'place makers' and 'place enhancers'.

Jones and Cloke (2002) emphasise so-called dwelling places, which are related to how humans are always embedded in the world; how they are, in other words, embedded in place. Dwelling is about the rich intimate ongoing togetherness of beings and things that makes up landscapes and places, and which binds together nature and culture over time. The concept of dwelling is potentially bound up with ideas of home and the local, and concern or affection for nature and the environment. Trees as place makers also help in defining dwelling places, as markers of time and representers of place, as well as the ongoing embedded interconnections of things and people mixing together in ways which mark each other and bind each other.

From the perspective of place identity and territoriality, one could imagine that city forests can act not only as public territories, but also as territories for a certain urban society that feels strong connections to a local forest.

The City Forest Defined

Thus city forests are cultural forest landscapes that are social and cultural constructs, created on at the meeting point of culture and nature, of the human and non-human. From an experiential perspective, they can be both space and place. But how do we define these city forests?

Obviously city forests are characterised by their close ties – economic, but also cultural and social, as we have seen – with a particular city. This close relationship is often the result of a long history, as will be illustrated throughout this book. Over time, citizens and municipal authorities have increased their influence in the decision-making about their local forests.

The close links and the joint development of the city as a community of citizens, with its own respective culture(s) comprising distinctive spiritual, material, intellectual and emotional features, and a local woodland, are key to the city forest concept. As a result of these ties, city forests developed a range of characteristics that set them aside from most other forests, such as their (mostly municipal) ownership, the prioritisation of social, cultural and environmental values, more designed character, high level of (recreational) infrastructure and their high diversity (Konijnendijk, 1999). City forests are also often small in scale and fragmented. 'Forest' in an urban context is often defined more broadly, with the inclusions of smaller woods, open areas and water. More 'modern' city forests such as the Amsterdamse Bos in The Netherlands, which has inspired the design of many of the recreational city forests that were established across Europe during the 20th century, have only a limited (e.g. between 30% and 50%) woodland cover.

Although they have close relations with cities, city forests are not city parks. City parks today are mostly seen as smaller, highly designed, inner-city areas primarily meant for recreational use. They include trees and even woods (i.e. smaller areas with trees as main vegetation), but mostly not large forested areas. Typically, their establishment and maintenance costs are much higher than those of city forests. Over time, however, the concept of 'park' has undergone changes starting as exclusive aristocratic hunting reserves – or "enclosed reserves for beasts of chase", as mentioned by Olwig (1996, p. 382). Gradually the park concept became extended to become a "large ornamental piece of ground, usually comprising woodland and pasture, attached to a surrounding house or mansion, and used for recreation, and often for keeping deer, cattle or sheep" (Olwig, 1996, p. 382). This is quite different from the municipal park of today. Muir (2005, p. 113) writes about parks: "[these are] places where scenery has been created and maintained in an idealised form. They are seen as places for spiritual enrichment and as refuges for the temporary evasion of the ugliness and clamour associated with the world outside their gates, and their potentials for recreation and contemplation are thought to have primacy over the stresses and tedium of the workaday world."

Not surprisingly in the light of this, the boundaries between city park and city forest are not always clear. 'Forest' has been used as referring to exclusive, often enclosed hunting areas, as has 'park'. In the wilderness, on the other hand (wild-deer-ness, as Olwig (1996) points out), beasts ran wild. The presence of trees was not a given for 'forest' nor 'park' in this original meaning. Modern examples of city forests show that the boundaries are still unclear. The Amsterdamse Bos is seen as a 'forest' (as suggested by its name), but it has all the characteristics of a (very large) park. Many city woodlands are small, only covering a few hectares. 'Park' today has primarily positive associations and is used in a whole range of contexts, such as in business park, science park and amusement park (e.g. Dings & Munk, 2006). The city forest of today is perhaps part park, part wilderness. Or, in the terminology of people like Tuan (2007), part place, part space.

The changing definition of 'forest' has already been discussed. Apart from the definitions given earlier, forest also has meaning on a deep symbolic level, as the 'other' to western civilisation, as a 'sylvan fringe of darkness' (Harrison, 1992; cf. Perlman, 1994). In the forest, humans are enclosed, dwarfed, concealed and/or lost. Forests have offered a retreat from modernity, providing both refuge and threat, symbolising both the noble savage and the backward and uncivilised savage (e.g. Schama, 1995). The term 'woodland' is often used interchangeably with forest, although it suggests a more intimate, more culturalised landscape (e.g. Jones & Cloke, 2002). Woodland is owned and managed, in contrast to the wildwood (Rackham, 2004). In this book, the terms 'forest' and 'woodland' are often used interchangeably. Forest and woodland both refer to wooded areas of a certain size, perceived as providing a 'forest feeling' or experience to visitors. 'Wood' is sometimes also used in this book, although this usually refer to a smaller wooded area (e.g. less than 1 ha) than forest or woodland.

Expert definitions of forest do not have to converge with the ideas of the ordinary forest user. Recent years have generated a number of studies that look at how people define, perceive and use forests in their (urban) environment. In-depth interviews with people living near Brøndbyskoven, a small city forest near Copenhagen, Denmark, showed that these people define 'forests' by the presence of trees, but also in terms of being closed or open, light or dark. Some respondents characterised the forest primarily as a place for people, others as a place for animals and plants. Interviewees saw the forest as a contrast to hurried life, as a place to 'disconnect' and to contemplate. For some, the forest acted as contrast to the civilised and cultivated. Most people preferred a limited amount of facilities as well as other people in the forest (Oustrup, 2007). Once again, the city forest as being both 'place' and 'space' becomes clear.

Contents of This Book

Human connections with natural resources and the landscapes in which they occur are multi-faceted, complex and saturated with meaning. There is a continuous construction and reconstruction of place and space through social and political processes. These processes assign meaning, in combination with biophysical attributes and processes, and are based on social and cultural values (Cheng et al., 2003). This book tells the story of one of our important landscapes, that of city forests, from a historical as well as modern-day perspective. City forests are studied as cultural landscapes, as places jointly defined by local communities and trees, and as space that contrasts with more controlled city environments. This book should not only help people to understand city forests as such, but also offer greater insight of the joint development of cities, society and nature. As stated by Jones and Cloke (2002, p. 1), the last 20 years of the 20th century "witnessed a very significant increase in the importance of understanding nature-society relations as an integral part of the political, economic, social and cultural constitution and reconstitution of changing places". Insight into the 'place making' role of city forests will also be valuable in the continuous quest to develop good cities.

The next two chapters of this book will look at the 'deeper' cultural links between people, culture and city forests. In Chapter 2 'The Forest of Fear', various human fears of the 'dark and wild' forest are explored. While forests were essential for the survival of ancient man, they also represented 'the other', the uncontrollable and wild. Some of the primeval fears for forest persist even today. Other fears relate to dangers caused by criminal behaviour of other humans. The chapter also shows, however, that fear of (city) forest can be turned into excitement as an antidote to the routines and control of urban life. Ancient cultural links between forests and people are also the topic of Chapter 3 'The Spiritual Forest'. Forests and trees have played important roles in religion and the forming of myths and symbolism that are at the foundation of cultures. City forests are also important in this respect, for example as early religious groves, as setting for myths and legends, as well as a basis for modern spirituality.

Chapters 4 and 5 take a more utilitarian view of city forests. Chapter 4 'The Fruitful Forest' shows how urban societies have used nearby forests as crucial source of products, and most notably timber. Cities were literally built on forests and wood, while city forests have also provided food, fodder, water and other products. As shown in Chapter 5 'The Forest of Power', not everybody had the same access to these forest products. Prevailing power relations meant that a limited few, starting with kings, aristocracy and church, could reserve certain user rights for themselves. The use of city forests as hunting areas, in particular, played a crucial role in this respect.

City forests, of course, have been appreciated for reasons other than providing wood and other products. As urban societies developed and cities expanded, appreciation for nature became renewed, offering an alternative to congested urban conditions and a nostalgic link to times long gone. City forests were an obvious place to 'escape' to and to live and recreate in more sedate, natural landscapes, as shown in Chapter 6 'The Great Escape'. Initially, the recreational use of city forests was a matter of the higher classes, who had sufficient time on their hands, as well as the means to create their own city forest escapes. But gradually city forests became recreational escapes for all segments of urban society. The re-appreciation of nature and forests was also reflected in art, as writers, painters, poets and others found their inspiration in city forests. Chapter 7 'A Work of Art' thus looks at the links between city forests and art as an important element of culture. Not only were artists inspired by city forest landscapes, they also played an active part in the first efforts to conserve the nearby natural environment. Chapter 7 also discusses city forests as 'works of art' by themselves, as their design and management have been 'coloured' by cultural changes, fashions and international exchange. Today, art can also be used as a tool for enhancing links between urban people, communities and city forests. The important role of city forests as a natural 'antidote' to cities is described in Chapter 8 'The Wild Side of Town'. City forests represent a more or less 'wild' nature, partly explaining their popularity as a recreational environment. They can help us keep in touch with nature and natural processes. Moreover, studies have shown that having some sort of 'wilderness' in or close to people's living environment fulfils important spiritual and psychological needs.

The latter links directly to Chapter 9 'The Healthy Forest', where city forests are regarded as 'therapeutic landscapes'. For quite some time, the contributions of city forests to human health and well-being have been known and recognised. However, in the light of major health challenges associated with modern, urban societies, such as obesity and stress, the positive roles of nature and city forests are once again in focus. Forests and nature have been found to improve mental as well as physical health and well-being in various ways. This important function offers an important direction for future city forestry, as does the theme discussed in Chapter 10 'The Forest of Learning'. Over time, city forests have been educational landscapes, for example acting as testing grounds for forestry. City forests, for example in the form of demonstration forests and landscape laboratories, have also improved the dialogue between foresters, other experts, and the public. Furthermore, city forests play a crucial role in environmental education, particularly of children and youths.

Chapter 11 'The Social Forest' considers yet another important socio-cultural role of city forests. These forests are an important element of and even contributor to 'place making', the development (or maintaining) of close links between local (urban) communities and their environment. City forests provide a meeting and assembly point, help provide identity and offer a way to strengthen community ties. This is important in a time of globalisation, urbanisation and individualisation, when local identity is not easily found. The sustainability of city forests can be threatened, however, by conflicts over these forests and their use. As Chapter 12 'A Forest of Conflict' shows, city forest conflicts are many and varied, from efforts to protect these forests from urban development, to fights between different types of recreational and other forest uses. From a more positive perspective, however, conflicts are a sign of the commitment of people and communities towards 'their' city forest.

In the final chapter of this book, Chapter 13 'Forests for the Future', some directions for the future of city forestry are provided. How can we strengthen and use the many social and cultural roles of city forests to create 'better' cities? How do city forests contribute to wider, multifunctional cultural landscapes in a rapidly changing world? But also: how do we develop the role of city forests as both 'place' and 'space', as a way to strengthen local communities and place identity, while also offering a 'wild' and exciting antidote to artificial, controlled urban life? After having read this book, the reader will hopefully agree that our city forest heritage is something to cherish, but also to develop.

Chapter 2
The Spiritual Forest

This chapter looks at the spiritual meanings of forests, and particularly of city forests, over time. The spiritual values of something do not only relate to religion, as explained by Schroeder (2001). 'Spiritual' refers to the experience of being related to, or in touch with, an 'other' that transcends one's individual sense of self and gives meaning to one's life at a deeper than intellectual level (Schroeder, 1992). A spiritual experience is an intuitive and emotional kind of experience in which a person feels caught up and carried along – or alternatively filled and inspired – by a feeling, an idea, an image or a creative impulse. This can be through religious rituals and disciplines, but also through contact with forests or other parts of the natural environment.

Forests have been described as spiritual landscapes (e.g. Jones & Cloke, 2002). As a refuge and source of subsistence, forests were very important in the spiritual life of the first humans, as is described in the first section of this chapter. The second section shows that religion has always turned to forests, as landscapes of inspiration, places of worship (as in the case of groves) or as a landscapes of income for religious institutions. The chapter's third section looks at forest myths and the symbolism of forests and trees, as these are part of what shapes a culture. Section four shows that 'forest spirituality' is still present today, for example, in the 'new heathen' communities that exist in several countries, as well as in the many spiritual links between people and the (city) forests they visit. In the final section, the use of the spiritual dimensions of city forests as a way to connect local communities and local forests is explored.

Ancient Spiritual Links Between Forests and People

The spiritual and symbolic values of forests and trees have a long history. The forest has been an important archetype, symbolically reflecting the basic structure of the unconscious, the undefinable, and the essence of life and existence (Schroeder, 1992; Koch, 1997). To our ancestors, forests were an all encompassing environment, even though the so-called prospect-refuge theory (Appleton, 1996; cf. Perlman, 1994) suggests that ancient humans preferred the forest edge and the

C.C. Konijnendijk, *The Forest and the City: The Cultural Landscape of Urban Woodland*

savannah, which offered a good overview for hunting as well as shelter, over the 'deep' forest. Forests were first, they were antecedents to the human world, offering the precondition or matrix for civilisation (Harrison, 1992). The clearing of the forest, however, made it possible to claim a dwelling where relatives could be buried. According to Harrison (1992, p. 7): "[b]urial guaranteed the full appropriation of ground and its ultimate sacralization." By burying their ancestors, humans started to root themselves quite literally in the soil (cf. Harrison, 2002). Thus a place of dwelling was created (Jones & Cloke, 2002). Time became more linear, gradually moving away from a primary focus on the seasonal cycles of nature (Gallagher, 1993; Tuan, 2007).

Civilisation developed where 'wild' forests were cleared. Harrison (1992, p. 11) cites Giambattista Vico's 'New Science', which first appeared in 1744: "This was the order of human institutions: first the forests, after that the huts, then the villages, next the cities, and finally the academies". Increased order was imposed on mankind's environment – order being systematic, progressive and self sustaining. All order is haunted by finitude, however, as great civilisations fell and forests once again took over. Harrison (1992, p. 12) puts this nicely into words: "At the end of the order of institutions cities become like forests in the figurative sense – places of spiritual solitude where savagery lurks in the hearts of men and women – but this demoralization merely prepares the way for a literal metamorphosis of the city itself. As the city disintegrates from within, the forests encroach from without." The Roman Empire, with its great capital city Rome, was among those civilisations that fell. In fact, Rome was 'defeated' by the forest in more than one sense, as its armies were ultimately overrun by German tribes of forest dwellers. Earlier, Roman armies had destroyed the forests that had been so central to the culture of these tribes. But Roman civilisation had an ambiguous relation to the forest, at it also originated from it. As a result, Roman culture included a nostalgic longing to these forest roots (Harrison, 1992; Paci, 2002).

Given the central role of forests in the rise and fall of civilisations, it is not surprising that ancient and sacred writings hold many references to forests (Porteous, 2002). Trees and forests feature prominently in most religions, as the next section will illustrate. In traditional cultures, trees have been associated with spirits, fairy folk and the like. Spiritual considerations played an important role in how people dealt with forests. In ancient times, for example, the state of the moon, i.e. whether it was waxing or waning, was considered to be of great importance in connection with the felling of trees (Porteous, 2002). Historical sources mention particular timber uses in relation to a specific felling date which supposedly ensured advantageous wood properties (Zürcher, 2004).

Over time, forests and nature became something that needed to be controlled and dominated by humans. Later still, when many of the world's forests had disappeared, a re-appreciation of nature occurred, as some of the following chapters will describe. Forests were regarded as the origin of humans and as antidote to corrupted society and to the all encompassing city. They were considered a spiritual retreat, a place to relive old connections and cultures.

Forest and Religion

Forests and religion have also been closely associated. Initially the forest 'obstructed' the connection between people and their gods. Later, it was the very same forest that provided the groves where gods could be worshipped. Where the city had shrines, the countryside had groves as sacred space (Tuan, 2007). Many of the first groves originated from sites on hills and mountains that were logical places of worship, as people were nearer to heaven (Porteous, 2002). Later, these sites were planted upon, providing tree dwellings for divinities. In this way, sacred groves emerged. The establishment and protection of groves have been a widespread tradition in many cultures. Groves were often associated with secrecy and initiation rites (Crews, 2003). Celtic druids, for example, have had groves as their sacred sites (Caldecott, 1993). Groves also feature prominently in religious texts such as the Bible, as in the case of the oak grove at Mamre. In Central and Northern Europe, dark groves composed of ancient trees and situated in the midst of gloomy forests acted as first temples. An invisible deity – or its spirit – dwelt in the groves, expressing its powers through storms and sunshine (cf. Perlin, 1989). Groves could seldom be entered by people other than priests and druids.

Groves could be found across Europe and also near towns. For example, sacred groves of hazel (*Corylus avellana* L.) formerly existed in the vicinity of Edinburgh and Glasgow, Scotland. Many of the remaining smaller woods in Germany are considered to have been groves, such as the Odenwald or Odinswald (the Forest of Odin). The legend goes that the city of Vienna arose around a sacred grove, the last remaining tree of which, a venerable oak (*Quercus*), gave its name to the 'Stock am Eisen Platz' in the centre of the city.

However, when pressures on forests increased with population growth, eyes also turned to the groves, in spite of their sacred status. Protective measures were thus needed. In Ancient Greece, for example, one of first decrees protecting sacred groves, taking effect during late 5th century BCE, came into force on the island of Kos (Perlin, 1989). It stipulated that anyone cutting down cypress trees would be fined 1,000 drachmae – about three years' pay for a worker at the time. From the 4th century BCE, deforestation in Ancient Greece intensified and laws to protect sacred groves became even more numerous. Later, the clearing of groves intensified, as the Christian Church wanted to get rid of groves and the 'heathen' practices they symbolised (Crews, 2003).

As mentioned, groves were often situated near cities, sometimes providing the basis for their very foundation. In other cases, groves were incorporated by expanding cities and towns. Not only groves were important to religion, but also individual trees. As many cities were established as sacred places (e.g. Tuan, 2007), some of the first trees in cities were those planted around temples during Greek and Roman times (Lawrence, 2006). Trees and plants in private and monastery gardens often carried symbolic meanings and cultural associations that were frequently religious in nature. The very concept of a garden, of course, has biblical associations with Paradise. As mentioned already by John Evelyn during the 17th century, rowan trees (*Sorbus aucuparia* L.) in Wales were very sacred and no churchyard was

Fig. 2.1 The sacred Bodhi tree (*Ficus religiosa* L.) near the ancient capital city of Anuradhapura, Sri Lanka attracts many thousands of Budhist worshippers (Photo by the author)

without one (Porteous, 2002). Trees also feature prominently in other religions. The Bo(dhi) tree (*Ficus religiosa* L.) at Anuradhapura, Sri Lanka is said to have been grown from the cutting of the original tree under which Buddha found enlightenment. It is also allegedly the oldest living tree with a written history, as writings identify the tree as originating from the 'fourth generation' of cuttings from a parent tree planted in the site in 236 BCE. The sacred Bodhi tree of Anuradhapura has been called a world heritage (Swarnasinghe, 2005; see Fig. 2.1).

In Europe and other parts of the world, Christian institutions have played an important role in city forest history. Monasteries did not only own many of the first urban trees, they also established and managed wooded estates near cities (e.g. Schulte & Schulte-van Wersch, 2006). Woodlands were part of large land possessions and estates that generated income for the Church. The Lorettowald in the

German city of Konstanz, for example, has been in the hands of monasteries and religious organisations during most of its over 800 years of history. Today, this 166 ha woodland is owned by the 'Spitalstiftung Konstanz', a hospital foundation with religious roots (City of Konstanz, 2007). Large areas of what are now the Bois de Vincennes and Bois de Boulogne were owned by monasteries, which also had traditional user right on other nearby woods (Derex, 1997a, b). Monasteries were also important actors in the Zoniënwoud near Brussels. Here, religious communities were granted special permission to use the forest and to settle in and build in it (van der Ben, 2000). During the Middle Ages, Epping Forest near London hosted various religious communities and an abbey. Many paths were probably made by pilgrims who were seeking comfort from the Holy Cross, held by Waltham Abbey (Green, 1996).

However, most of the woodlands owned by the Church were later confiscated by the Crown or State. Wiggins (1986) describes how many of the woodlands in what is now the Telford, United Kingdom area were monastic lands until the Dissolution of Monasteries (cf. Green, 1996). Several forests near Tallinn (then known as Reval), Estonia, were owned by religious communities, but these church and monastery forests were expropriated by the state during the 16th century Reformation period (Meikar & Sander, 2000). One of Tallinn's present city forests is still named 'Klosterwald'. The so-called 'Kapucijnenbos' in the Zoniënwoud, which was named after the order of Capuchin monks, was sold to a royal foundation and turned into the Tervuren Arboretum (Van Kerckhove & Zwaenepoel, 1994; van der Ben, 2000). In Sweden, Norra Djurgården near the centre of Stockholm was formerly owned by a monastery and transferred to the royal family only during the 16th century (Kardell, 1998).

City forests have also included religious elements and offered religious inspiration (Fig. 2.2). 'Healing springs' acted as points of pilgrimage. Examples of these include the 'Kirsten Pils Kilde' in Jægersborg Dyrehave north of Copenhagen (Møller, 1990; Olwig, 1996) and the 'Uggelvikskällan' in Norre Djurgården (Kardell, 1998). So many people came to Kirsten Pils Kilde that it became the obvious place to establish the still existing 'Bakken' amusement park. The forest of the town of the Polish town of Wejherowo hosts 26 religious chapels and shrines along a pilgrims road. The use of the forest by pilgrims and the resulting compaction of the soil have sometimes been so high that foresters have experienced difficulties with regeneration of forest stands (Konijnendijk, 1999). Similar roads of pilgrimage, often leading up to a chapel or a temple and thus mimicking to some extent the Via Dolorosa (the last walk of Jesus Christ), can be found in other city forests. Olmsted Woods, the site of the Washington National Cathedral, for example, includes such an uphill 'pilgrim way' (McIntyre, 2006).

City forests provide natural, spiritual settings for churches, other places of worship and cemeteries. Forests have been compared with cathedrals and the long history of sacred groves and the symbolic values of trees, connecting heaven and earth, life and death, all make this link obvious. The mentioned Olmsted Woods, a 2 ha woodland in Washington D.C., United States, make the approach to the cathedral up on a wooded hillside more dramatic. The approach

Fig. 2.2 Many city forests still host religious elements. The original of this little Maria-chapel in the Zoniënwoud of Brussels was installed by Queen Maria-Henriette (1836–1902). Today a replica remains, mounted on the original beech tree (Photo by the author)

to the cathedral, seemingly 'floating above the trees', helps to forget the hustle of daily life and stimulates contemplation and cleansing of mind and spirit, as mentioned by McIntyre (2006).

Some countries, perhaps most notably Sweden, have established 'forest cemeteries'. Stockholm's 'Skogskyrkogården', which combines trees, mounds, open spaces and, of course, graves, is a UNESCO World Heritage Site (Nilsson, 2006b). Sosnovka woodland park in St. Petersburg, Russia, hosts a cemetery for soldiers fallen during the long siege of the city – then called Leningrad – during World War II. Jones and Cloke (2002) describe how the Victorian-era Arnos Vale Cemetery in Bristol, UK, gradually developed into a sort of woodland park with both planted (often exotic) trees and spontaneous regeneration. Trees and woodland elements have always played an important role in cemeteries, symbolising life, death and eternity, as well as contributing to an appropriately serene and sacred atmosphere through their texture, colour, shape and sounds. Threes have also been described as 'sentinels to the seasons' (Jones & Cloke, 2002).

During modern times, the role of cemeteries has changed. Cemeteries in the densely-built heart of Copenhagen, Denmark, for example, are currently also very popular areas for outdoor recreation, acting as de-facto city parks (Guldager, 2007). In general, the role of cemeteries is under debate (Guldager, 2007; Worpole & Rugg, 2007). Although churchyards or local cemeteries have always been a key symbolic space in the history of settlements of all sizes, Worpole and Rugg (2007) argue, one can question whether so much space should be reserved for burials. In Britain, for example, currently about 70% of the deceased are cremated. On the other hand, there is a desperate need for places of memorialisation in modern society, as can be seen by the increase in the number of roadside memorials. Maybe this opens the way for other kinds of places of remembrance, such as small memorial parks or dedicated sites in city forests and other green spaces (cf. Guldager, 2007). Debates are taking place in countries such as Denmark, The Netherlands and the United Kingdom about extending possibilities for being buried – or at least for dispersing the ashes – in nature. In The Netherlands, for example, several 'nature cemeteries' have been set up where people are buried in an 'environmentally-friendly way' (in a bio-degradable coffin), often at the foot of a tree (Nijen Twilhaar, 2007). In the Scottish city of Edinburgh, several city woodlands have been allocated as green burial sites (Woodland is earmarked, 2004).

A case from Malaysia reported by Reuters India (Fernandes, 2007) also illustrates that the relationships between religion and city forests have changed. In the year 2005, a conflict arose over the establishment of a new Muslim cemetery in the wooded area of Sungai Buloh near Kuala Lumpur. Local officials said that the cemetery would occupy just a small part of the forest, but residents feared that the move was only the first step in a wider effort to cut down trees, level the hill on which they stand and build hundreds of homes on the site. The woodland had already lost a large part of its area during the preceding years. One local dweller and activist preparing to take on developers and town planners stated: "why dig up a valuable tree to bury a corpse?"

Forest and Myths

A symbol is a word or image that stands for something other than itself. Symbolism allows a concrete object (a tree, for example) to represent an idea or experience that is intangible, indefinite, or only vaguely understood. Kockelkoren (1997) writes about the modern symbolism of forests, explaining how forests have become cultural-historical icons. These icons do not only directly depict something, but they also embody it. According to Kockelkoren, forest icons refer to the Paradise (the lost Garden of Eden, the lost harmony between pure nature and man), the Home of the Noble Wild (the tension with culture that started clearing the forest, the opposition between culture and nature) and the Wood of Robin Hood (a forest

of freedom, of joint fighting against the rules of law). The latter can be recognised in the English 'Greenwood' ideal as described by Schama (1995), Nail (2008) and others (see the Chapter 11 'The Social Forest').

Cultural-historical icons are also the matter of mythology, which uses symbolism to express the basic beliefs and values of a whole culture (Schroeder, 2001). A myth can be described as any real or fictional story that appeals to the consciousness of a people by embodying its cultural ideas or by giving expression to deep, commonly held emotions. With myths, storytelling is used to overcome fear of the immensity of the university and the mystery of our being here (Caldecott, 1993). As Tuan (2007, p. 85) states: myths "flourish in the absence of precise knowledge". Myths may be based on real, historical events. They can also provide an important source of insight into the psychological and cultural relationships between people and natural environments. In fact, as mentioned by Jacobs (2004), many myths in Indo-European cultures are very similar in structure. As it is reasonable to regard myths as reflecting the fundamental structures of a culture, the similar myth structures relate to the fundamental principles in a culture.

Forests and trees feature prominently in mythology and thus in shaping cultures. Trees have often been symbolising the fundamental values of life, for example, as expressed in the myths of the Tree of Life and the World Tree. Trees have long been symbols for all manner of key social meanings and practices (Perlman, 1994; Jones & Cloke, 2002; Crews, 2003; Muir, 2005). Simon (2001), for example, mentions how in Ireland, trees are linked to a wealth of traditions and folklore. Even the ancient Gaelic (so-called Ogham) alphabet was based on the names of trees and other woody plants (Caldecott, 1993). Trees have been identified with human form, symbolise life cycles and continuity and connect order and chaos. They link the present world with the spiritual world, a union of earth and heaven. The forest is the 'other', representing the 'otherness' of living beings other than human (Caldecott, 1993; Perlman, 1994; Crews, 2003). It often symbolises that which cannot be rationalised or fully understood (Harrison, 1992). Contrasts between the young and the old, the native and the alien, summer and winter and so forth have also been symbolised by trees as well as forests. During May Day rituals of renewal, initiation and fertility, village communities and youths went into the woods and returned with maypoles (Muir, 2005; see Fig. 2.3). According to Jones and Cloke (2002), the older a tree is, the more likely it is to become the focus of particular attention and value.

Porteous (2002) discusses many of the symbolic values of trees and their featuring in myths. Trees have presided over marriages, for example, and often trees are found which represent the betrothed and afterwards the young married couples. Trees are also planted at childbirth, even up until today – something which is used, for example, in the establishment of new city forests. The former president of Iceland, Vigdis Finnbogadóttir, used to plant trees all over her country, at different official occasions. She always planted three trees of the same species: one for the children of today, one for the children who had died, and one for the children yet to come (Kissane, 1998; E. Ritter, 2007, personal communication). Trees grow with the children and symbolise health and offspring, leading them to be carefully tended. As mentioned, trees are also linked to death. The funeral trees of Naples are

Fig. 2.3 The maypole, symbolising a tree, is still an important element of Midsummer Eve celebrations in Sweden (Photo by the author)

sombre and dark trees to which a fatal power is attributed. Other symbolic trees include the Christmas tree and mythical trees such as the mentioned World Tree, the ash (*Alnus* spp.) Yggdrasil from Nordic mythology and the Tree of the Cross. Sometimes individual trees are attributed with special power or symbolic (and often religious) value, such as the Glastonbury Thorn (hawthorn, *Crataegus monogyna* Jacq.) which was thought to have sprung up from the walking staff of Joseph of Arimathea (cf. Muir, 2005).

Legendary stories have become associated with trees and forests. Some of these stories made a direct link between cities and forests. The ancient epic of Gilgamesh, for example, tells how a king from the city of Uruk took to the nearby Cedar Mountain to slay the mythical forest guardian, Huwawa. It is a story of the struggle between city and forest, between nature and culture, and of a man's quest for immortality (Harrison, 1992; Paci, 2002). Myths and epics often speak of the ambiguous relations between humans and the dark forest. Examples are the stories about Herne the Hunter, one of the keepers of Windsor Forest during the reign of Richard II, and the Finnish epos of the Kalevala (Porteous, 2002).

Tales of Herne, by the way, are a recurring theme in myths across Europe. In France, the wild hunter is called 'le grand veneur', for example (Porteous, 2002). This shows, as mentioned before, that there are some recurring stories, themes and characters in forest-related mythology, in spite of local differences. Certain characters feature in the forest-based myths and folklore of regions and societies in different parts of the world. Witches or old women living in the forest feature in Swedish tales of the Vargamor or Wolf-crones, as well as in stories of the Wood Wives (Holzfrau) as captured by the German Grimm Brothers. Many myths and fairy tales depict woods as dangerous places and offer words of caution. But forests have many levels of meaning. In boreal fairy stories, forests symbolise a setting where 'inner darkness' is confronted, where uncertainty is resolved and where people start to understand who they want to be (Clifford, 1994).

Forests have also helped preserve ancient, local culture, as discussed by Harrison (1992). At least until medieval times and in some countries until much more recently, each city and town was rather isolated and centred towards itself, as an island in the 'sea' of vast forests (or wilderness) surrounding it. This relative isolation meant that local myths and folklore could develop and also survive, without too much interference from other areas.

During the time of Romanticism, which started in the late 18th century, many of the 'old' symbolic values of trees and forests were rediscovered. Trees were once again regarded as powerful symbols of nature (Lawrence, 1993), while trees and woods were also seen as symbols of wealth, continuity and sustainability of the family line, especially by the upper, land-owning classes (Irniger, 1991). Folklore was rediscovered as an instrument to get back to ancient knowledge and to the roots of civilisation. The previously mentioned German Grimm Brothers, for example, compiled a collection of Central European fairy tales that was published at the beginning of the 19th century. Folklore acts as a bridge between nature/forest and people (e.g. Harrison, 1992; Lehmann, 1999). Forest myths, folklore and sagas are of high importance for a culture's collective memory (Lehmann, 1999). Moreover, folktales and myths are an important part of the cultural process, assisting in adapting society's norms and joint history. Many folktales and myths deal with the ancient struggle between the traditional subsistence rights of the common population and the monopoly of rulers (primarily relating to hunting rights). 'Evil' foresters patrolling the woods and punishing offenders in the name of the land owner were seen as the 'bad guys'. The so-called 'Oberwächter' (head guardian) of Kaiser Maximillian II of Austria, who was ordered with the task to prevent poaching in the Prater hunting domain of Vienna, was only one infamous example of this (Pemmer & Lackner, 1974).

The mystic role of trees and woods is still present, in spite of the emergence of rationalisation and scientific forestry and the destruction of forests, myths and ancient culture. In some cases, myths are still reflected in the actual (city) forest. For example, the 'Rad', a wheel-shaped site in a clearing of the Hanover Eilenriede forest, symbolises the rituals of old Germanic times, when people took to the forest to dance in circles and engage in spring-time games (Fachbereich Umwelt und Stadtgrün, 2004). The 'Hexenhäusl' (small house of the witch) in the Forstenrieder Park near Munich was allegedly a place where ghosts resided, but it is now in use by the forest service (Ammer et al., 1999). Myths are also still reflected in rituals such as those of tree planting ceremonies and the carving of names in tree bark. The history of a forest or tree is sometimes regarded as part of a personal history (Perlman, 1994). In the United Kingdom, particular trees are valued for various reasons and venerated by many as icons of longevity, stability and permanence. In the United States, natural features including parks, gardens and trees have been found to stand for endurance and local distinctiveness (Jorgensen & Anthopoulou, 2007). Woodlands evoke memories of childhood and also give rise to the dream of rural living, or of an imagined arcadia.

New myths are continuously being created, for example about criminals hiding in the forest (see Chapter 3 'The Forest of Fear'). Another example is the story of the 'world's largest Swastika forest' near the German town of Wiesbaden-Naurod. Allegedly the forest at this site had been cut and shaped in the form of a swastika during the era of national-socialism. This shape would still have been recognisable long after the war. The story of these and other swastika forests, however, are now generally considered to be the material of 'urban legends' (Lehmann, 1999).

Modern Spirituality

The spirituality of the forest was 'rediscovered' as early as the 19th century by groups such as the 'Wandervogel' ('walking birds'). In Austria, these groups took long walks in the surrounding hills and forests of the 'mystical' countryside near Vienna and other cities, trying to find a tribal or racial 'Heimat', and to regain their health lost in the corrupted city (Rotenberg, 1995). Others tried to overcome the corruptions of the city and find nature in homeopathy and nudism. As seen, forests and trees continue to have spiritual importance also today. The present-day attitude of people to living trees is often sentimental and romantic. Trees, unlike most other plants, are granted a soul (Perlman, 1994; Lehmann, 1999). Trees and woods link us with the past and with rural roots, as a study of elderly people in the UK and their relationship with woodland confirms (Jorgensen & Anthopoulou, 2007).

The spiritual roles of (urban) woodland today are confirmed by many authors, including Kaplan and Kaplan (1989), Coles and Bussey (2000) and Jorgensen et al. (2006). People's encounters with nature seem frequently accompanied by existential ruminations. Human-nature relations provide a window on how people cope with matters of life and death (Koole & Van den Berg, 2004). Pirnat (2005; cf. Perlman, 1994) suggests that human life cycles can be linked to those of forests and trees, with the newborn equalling the sapling, the elderly being associated with ancient trees, and so forth. As life and death have increasingly been placed 'outside' the family home (i.e. in hospitals, nursing homes), forests can help to keep people

Fig. 2.4 The nearby forest, as here in the case of a woodland in Holstebro, Denmark, can provide a setting for contemplation (Photo by Anders B. Nielsen)

in touch with the circle of life, although tree lives are generally longer than those of human beings.

Human-nature relations today are complex. Close encounters with nature involve confrontation with deeply rooted existential fears, which fuel defensive motives to distance oneself from, or try to control, the wild forces of nature (see Chapter 3 'The Forest of Fear'). However, nature also provides an ideal setting for exploration and personal growth (Fig. 2.4). A study among members of 20 Dutch nature organisations showed that being in the woods was associated with inner peace, the experience of connectedness and reflections on the cycle of life and death and one's smallness in the grand scheme of things. This shows a depth beyond 'just' recreational experience, often primarily oriented towards the pursuit of brief, care free pleasures and the need to 'recharge the batteries' (Cohen, 1979, cited by Koole & Van den Berg, 2004).

'Wilder' nature plays a particularly important role in terms of spirituality, offering 'space' for freedom and adventure in contrast to safe 'place'. Wilderness experience may provide the stress-reducing and attention-restoring benefits of everyday nature in a more long-lasting way (e.g. Kaplan & Kaplan, 1989; Gallagher, 1993). Wilderness experience has also been associated with a range of spiritual and transcendent experiences that provide benefits such as greater self-confidence, sense of belonging to something greater than oneself, and renewed clarity on 'what really matters' (Knecht, 2004).

A study by Williams and Harvey (2001) of 131 Australians who visit, work or live in forests illustrates that so-called transcendent experiences are associated with forests. According to Laski (1961, cited by Williams & Harvey, 2001), transcendent experiences are often triggered by nature or wilderness experience. Characteristics of these, as confirmed by the Australian study, are strong positive affect; feelings of overcoming the limits of everyday life; a sense of union with the universe or some other power or entity; absorption in and significance of the moment; and sense of timelessness.

Studies like these argue for the presence of 'wildwoods' close to where people live, as a supplement to the everyday nature of neighbourhood trees, gardens and parks. More about this can be read in Chapter 8 'The Wild Side of Town' and Chapter 9 'The Healthy Forest'.

Forests are an important setting and a source of mythical and religious rites and stories. Today, the terror of the forest seems mostly subdued, although Chapter 3 'The Forest of Fear' shows that feelings of fear are sometimes still present. The wild forest and 'real' nature are sought, in order to be associated with 'primeval' values. The primeval character of nature can also be found, for example, in pitch-dark nights which have become a rare phenomenon in the most urbanised parts of our world (Langers et al., 2005). Forests, also city forests, offer experiences related to fascination, novelty and compatibility. They provide privacy for reflective thought (Hammitt, 2002). In this respect, a familiar environment is considered with more affection or feelings of belonging (e.g. Williams & Harvey, 2001).

Ardoin (2004) states that people's interaction with places like city forests can be grouped according to different dimensions. Psychological elements feature in place identification and place dependence of an individual. As described in the Introduction, sense of place relates to an individualised meaning or fit between meanings attached to a physical place or a person's self image (Talen, 1998). Finally, socio-cultural elements also play a role in terms of dealing with the influence of the social community as well as the cultural context.

'Forest spirituality' is also reflected in the cults of 'new heathens' that search for spiritual enrichment amongst, for example, the trees of their local forest. Lehmann (1999) describes such a 'new heathen' community in Berlin, Germany. Its members see humans as integral part of nature, have animistic images of reality – i.e. elements of nature are living beings with a soul – and believe that trees have spirits. Lehmann estimates that there are between 5,000 and 50,000 of these 'new heathens' in Germany. They often have high forest awareness and fall back on myths, sagas and fairy tales.

The Danish religious community of Forn Siðr worships the old Nordic powers. It is based on advance knowledge of the old Nordic religion and wants to maintain the Nordic cultural heritage. It considers humans, animals, nature, holy places as hosting both life and spirit. On its web site (Forn Siðr, 2007) the community writes: "In Forn Siðr we have agreed on a basis of faith; this describes how we interpret Asatru in Forn Siðr. (…) The Asatru faith is growing strongly. Our gods, jætter, dwarfs, elves, and trolls are familiar concepts for most Scandinavians. Many Pagan elements have survived in Folk belief. Asatru is more than just this however, a tribute to both the Sun and the Moon, gratitude to the spirits of the land, and fellowship with nature, on which we all depend, and not least a way of life that is responsible to oneself and one's fellow humans." In November 2003, Forn Siðr was official recognised as a religious community by the Danish government.

Recreating Spiritual Links

According to Clifford (1994, p. 17), we "have to recapture and progress our long understanding of trees and the land, to re-enfranchise city people and those marooned in over-farmed countryside, to nourishing the imagination." Spirituality in its different forms can be an important aspect of relationships between people and city forests, for example, when new city forests are being developed. Kowarik (2005) mentions the case of new woodlands in the German Ruhr area. These new forests on abandoned urban-industrial sites have to overcome that 'the forest' is not held in high regard by local communities, as there are hardly any traditional ties between people and forests. What many local people perceive is a decline of the former economic structure, being overwhelmed by nature. New, 'deeper' links and place identity will thus have to be created. Experiences from elsewhere show that bonds can be strengthened. People engaged in the Forest of Belfast programme in Northern Ireland, for example, stressed the need to link to the wealth of Irish tree-related folklore and traditions for encouraging people to discover stories about local trees. In this way, local residents can gain deeper understanding of their surroundings (Simon, 2001).

In a New Zealand study of values people associate with vegetation in urban environments, psychological values as well as values related to culture, history and identity featured prominently. Vegetation in the living environment was associated with feelings such as good, peace, calm, harmony, restfulness, privacy and space. Trees in particular were regarded as providing continuity and a feeling of permanence. Vegetation was seen as influencing the identity a city creates for itself and how people feel about an urban area. On a spiritual level, urban vegetation offered an opportunity to remain in touch with nature. It was regarded as representing a living connection that people did not entirely understand (Kilvington & Wilkinson, 1999). This example illustrates the opportunities that exist in tapping into spiritual links between people and city trees and forests.

City forestry holds quite a number of examples of how spiritual links between people and forests are used. The Hamburg Environment Department established, in line with an old tradition, a Wedding Forest where trees can be planted at the occasion of a wedding, a baptism, and so forth (Lehmann, 1999). The previously introduced Forest of Belfast set up a Family Tree Site where families can sponsor tree planting to mark special family events (Simon, 2001). So-called 'magical woodland celebrations' have been held at Belvoir Park Forest, one of the components of the Forest of Belfast. Children and other visitors have been made familiar with the Forest by linking to history, involving them in traditional woodcrafts, and the like

(Woodland Trust, 2006). Also in the Forest of Belfast programme, the Lord Mayor of Belfast initiated the planting of a tree for every child born during the Millennium year.

Elsewhere in Europe, a visitor centre run by the Flemish government uses a fairy tale based on the Wagner opera 'Der Ring des Nibelungen' in an educational game for school children. A saga is used to introduce knowledge about forests and nature (Agentschap voor Bos en Natuur, 2006). A Dutch nature association follows a similar approach, organising special 'witch excursions' to its nature areas. Daughters and mothers can learn how to fly on a real broomstick, make magic potion and discover popular witch hideouts in nature (Natuurbehoud, 2006). As written earlier, in efforts like these, new myths are being created and new fairy tales are being written according to the demands of modern times. Also in The Netherlands, the traditional fairy tales presented in a popular amusement park have got out of fashion and are replaced by popular television characters such as Bob the Builder, as many children do no longer know the traditional fairy tales (Er was eens, 2006).

Religious organisations also continue their involvement in city forestry, be it less dominant than before. A local Christian group called 'Churches Together' partly funded the establishment of a new, 10 ha peri-urban wood at Cowpen Bewley in the Tees agglomeration of Northern England. The wood was given the name 'Faith Wood' to show the links between the local religious community and the forest. Planting in 1995 was inaugurated by a procession of Churches Together to the site (Konijnendijk, 1998). In Northern Ireland, the Forest of Belfast programme offered Irish yew trees (*Taxus baccata* L.) to churches of different denominations, to be planted at church sites (Simon, 2001).

At the beginning of this chapter, the issue of cyclical versus linear time was discussed. Treib (2002) also describes the opposing perspectives of cyclical and linear time, the first being closely associated with the rhythms of nature and the changing of the seasons. In Japan, the year traditionally centres on the passage of seasons. Festivals, floral displays, food and so forth all reflect the passage of time. The passing of the seasons, however, is not always that clearly perceived in cities. Therefore, green spaces and trees are crucial for marking the passage of seasons, apart from maintaining an overall connection with nature (Treib, 2002).

Spirituality taps into some of the deepest relations between people and forests. It can be a powerful tool in promoting community involvement in city forest planning, management and use. However, as the examples in this chapter have shown, The Spiritual Forest is a complex forest, where ties between people, woodland and elements such as trees are often highly individual. It appears in many shapes, changing with person and time. Thus generalised approaches are bound to fail, especially if they lack understanding of the local and individual spiritual context.

Chapter 3
The Forest of Fear

Generally, city forests are popular places, as most other chapters of this book will show. Yet, like all forests, they also have a 'darker' side. Jones and Cloke (2002) mention that 'places of trees' can be places of fear as well as of exclusion. In his account of his childhood in Des Moines, Iowa, author Bill Bryson (2007, p. 159) characterises local woods as follows: "The woods were unnerving. The air was thicker in there, more stifling, the noises different. You could go into the woods and not come out again. One certainly never considered them as thoroughfare. They were too vast for that."

This chapter focuses on the different fears that have been associated with forests, as fear has been an important element of ancient links between people and forests. Fear is an important part of landscapes and of the tension between place (home, safety) and space (the unknown, the adventurous and maybe unsafe). Jones and Cloke (2002) mention two strands of fear associated with forests and woods: the fear of what woodlands might contain or conceal, such as criminals, dangerous animals or mythical beings, and a 'deeper' fear, for example reflected in a fear of getting lost. At the outset of this chapter, the latter, 'primeval' fears of forests are dealt with.

In their account of safety and perceived safety in Dutch forests and nature, Van Winsum-Westra and De Boer (2004) distinguish between social and physical (un)safety. Partly based on this distinction, sections of this chapter consider forests as venues for criminal behaviour, as well as in terms of dangers associated with 'wild' nature. Finally, the often narrow divide between fear and excitement is used to demonstrate how forest fears can be managed and even used for the better.

Primeval Forest Fear

People have always had a love-hate relationship to forests. From the early days, forests provided humans with essential shelter, food, materials and inspiration. However, forests have also been associated with different kinds of threats. Schama (1995, p. 517) writes about the different and frequently ambiguous social meanings associated with forests. They represent two kinds of arcadia, "shaggy and smooth, dark and light, a place of bucolic leisure and a place of primitive panic".

In his book about the forest in folklore and mythology first published in 1928, Porteous (2002) introduces the rich heritage of folktales and myths related to forests. Many of the ancient myths and tales held a note of warning and were associated with people's fears. In the chapter on 'mythical denizens' of forests and woods, Porteous mentions that forests and woods have always been imbued with a certain degree of mystery. They have been believed, from early days on, to be peopled with "crowds of strange beings endowed with superhuman powers and characters, although partaking human form" (Porteous, 2002, p. 84). Some of these beings were benevolent, whilst others were malevolent. They included fairies, elves, witches, wood spirits and wood trolls, as shown in Chapter 2 'The Spiritual Forest'. Myths were – and sometimes are still – often used to 'rationalise' some of the workings of nature that could not be explained and therefore evoked feelings of fear. These 'workings of nature' included sounds made by the cracking of branches, lights of the fireflies at night, the night call of owls and other phenomena that evoked feelings of fear. Myths often held a note of warning, evoking fear in order to let people refrain from undesired behaviour, as in the case of Millington's 'Babes in the Wood' from 1595 (Muir, 2005).

Dwyer et al. (1991) describe how many people's fear of forest is connected to the past. In ancient times, the forest was a barrier for settling. It often harboured danger. Forests hosted large animals and were a possible hiding place for criminals. On the other hand, of course, they also offered a refuge to those perhaps unrightfully persecuted, as Chapter 6 'The Great Escape' will describe. People's fears of forests have been associated with the living environment of ancient man (e.g. Porteous, 2002). The prospect-refuge theory of Appleton (1996) suggests that our ancestors preferred the half-open landscape of the savannah where they could hunt for prey. The forest edge was also important, however, as it provided the necessary refuge. The deep forest harboured dangers on the one hand, whilst also providing some necessary shelter and a hiding place. Lehmann (1999) explains the association of forests and fear from the perspective of the forest being a dark border area, lurking at the fringe of civilised society. The forest edge, the author explains, is very important here, as entrance to the unknown. Interviews with German forest users – many of whom are living cities and towns and use nearby forests – indicated that people often have similar fears of being left behind in the forest, for example, based on childhood memories. This fear, like the fear of getting lost (Fig. 3.1), is a recurring theme. Many of the fairy tales and myths that have survived, and as for example written down by the Grimm Brothers in 19th century Germany, include a warning against wandering off (alone) into the dark and dangerous forest. The fairy tale of Little Red Riding Hood is an example of this. Many local stories with similar messages of warning exist locally. A returning fear is also that of a 'man behind the tree' (Lehmann, 1999). Lehmann's interviews with selected Germans indicate that quite a number of forest users regard every person they meet in the forest as a potential threat.

The primeval fears discussed above relate to forests at large, but they are also relevant to city forests and their use. Many people primarily come in contact with

Fig. 3.1 The fear of getting lost in the forest is a recurring theme (Photo by Moshe Shaler)

forest landscapes by visiting their local city forest. It is estimated, for example, that more than half of all forest visits in Sweden are to city forests (Rydberg, 1998). A recent study of urban woodland use in the United Kingdom illustrates that primeval fears still play a role (Jorgensen & Anthopoulou, 2007). Close physical engagement with wilderness-like environments, including forests, were found to inspire thoughts of freedom, but also of death amongst woodland users. Woodlands often contain highly enclosed environments, at least as compared to other landscapes, paradoxically inspiring feelings of both containment and freedom. The very enclosure, seclusion and other-worldliness that underpin the positive aspects of woodland (see Chapter 6 'The Great Escape' and Chapter 8 'The Wild Side of Town'), are at the same time also at the root of the fear and uneasiness that people experience when coming into contact with it. However, city forests are often more open and less dense than other forests. Partly this is due to their long-standing role as recreational landscapes, offering a mixture of woodland, water and open areas for different activities. Another reason for their more open structure is that design and management of city forests often take safety issues into account. This is reflected, for example, in the removal undergrowth and in improved visibility.

Criminal Forests

Although there is no evidence that forests are particularly dangerous places to be, over time they have become associated with various forms of crime. The British landscape architect and urban forester Alan Simson once wrote that it sometimes seems to be that every rape, attack and other crime reported in the British tabloids has at least some association with a wood (Simson, 2001a; cf. Milligan & Bingley, 2007). Lehmann (1999) also describes how forests have often been associated with (violent) crime, not in the least by the media.

Many forest fears are not caused by the forest itself, but rather by criminal behaviour of other human beings who use the forest with its secludedness and relative remoteness as a cover. When people are asked about the aspects of big cities they dislike the most, crime tops the list (Gallagher, 1993). Like other urban open space, urban woodland is a mirror of urban society and thus also features in debates about crime and crime prevention. Urban woodland offers a vast range of positive experiences to urban dwellers, but it also harbours less desirable activities. Woodlands and other green spaces in and near cities have always provided a less-controlled, 'fringe' environment for those who wanted to escape from the rule of law. As Dings and Munk (2006) explain, all those who want to embark on illegal or semi-legal activities turn to the city park (or in this case: the city forest), where there is less social control.

City forest history includes many examples of 'criminal associations'. When mass recreation started at places such as Epping Forest and Hampstead Heath near London, this also led to excesses and criminal behaviour such as prostitution, assaults and thefts (e.g. A history of, 1989; Muir, 2005). The annual Easter Hunt in Epping Forest started when the citizens of London obtained hunting and other rights to the forest. This event, which lasted until the mid-18th century, became an "occasion for drunkenness and rowdyism" according to Muir (2005, p. 161). Husson (1995) states that certain French forests have obtained a very bad reputation as places of exclusion, marginalisation and even violence. The Haagse Bos of The Hague, The Netherlands, had also gained a rather negative reputation by the end of the 20th century, although mostly amongst non-visitors. The area had become associated with drugs, prostitution and the presence of homeless people. More than one-third of the inhabitants of the city said that they did not feel entirely safe in the Haagse Bos (Anema, 1999). The presence of groups of youths in urban woodlands has also caused feelings of fear, as explained by Bell et al. (2004) in the case of Scotland.

Wing and Tynon (2006) carried out a crime mapping study in the national forests in the States of Oregon and Washington, USA. The authors analysed a database containing 45,000 spatially referenced crimes occurring in 2003 and 2004 on United States Forest Service land, from felonies to misdemeanors. The number of 45,000 crimes in two years sounds high, but one has to take into account that the total area studied comprises more than 115,000 km^2. The study's results indicate that crime densities do tend to be concentrated – perhaps not surprisingly – in forests adjacent to population centres and transportation corridors. It has to be noted, however, that the higher reported crime rates near population centres may partly be explained by more reporting of crime because of easier patrol access. Apart from crimes such as assault and drug activity, the researchers mention a group of 'urban-associated crimes', which include arson, body dumping, gang activity, rape and sexual assault, drive-by-shootings and murder. However, serious crimes such as murder and sexual assault do not occur very frequently, as nine out of ten reported crimes fall under the category 'misdemeanor', i.e. rather minor offences such as petty theft, vandalism and prostitution.

Police operating in the state forests of Wales give a similar impression of 'typical' crimes in the forest. The list of most common offences is topped by off-road trespass on motorcycles and 4 × 4s, criminal damage (primarily vandalism), fly tipping and auto crime (Forestry Commission Wales, 2007). Criminal damage also tops the crime statistics for the British Forest of Dean, a popular recreation landscape for near the city of Gloucester (Forest of Dean Crime and Disorder Reduction Partnership, 2005).

Although crime does occur in city forests and other green spaces, studies indicate that it occurs less frequent here than in other parts of the city. Crime mapping in the city of Gainesville, Florida, for example, showed that most parks were not really troubled by crime. A higher incidence of crime occurred in areas away from the city parks (Suau & Confer, 2005).

It cannot be denied, however, that serious crime is sometimes connected to city forests. There have even been cases related to terrorism. The terrorists of the German Baden-Meinhof group hid their arms in forests near cities such as Bremen and Hamburg (the Sachsenwald) and used these and other woods as hideouts (Lehmann, 1999). In a recent British case, a bag with elements to build a bomb was found in Kings Wood near High Wycombe, in the aftermath of foiled plot to bomb nine or ten transatlantic airliners (Terror detectives find, 2006).

Lehmann (1999) also describes how German forests – and often city forests – have been hideouts for bands of robbers and serial killers for centuries and up until today. One historical example is that of robber and murderer Jaspar Hanebuth, who made the Eilenriede forest of Hanover unsafe during the first part of the 17th century. Times were confusing, with wars and pest epidemics, leading to less law and order. Hanebuth benefited from these circumstances. He preferred the edge of the forest for attacking his victims. When Hanebuth was caught in 1652, he confessed to committing 19 murders (Fachbereich Umwelt und Stadtgrün, 2004). A recent case is that of Findeisen-Fabeyer, the 'forest man'. This murderer of a policeman hid in a forest near the city of Osnabrück until being caught in 1967. His crimes were blown up to legendary fame and resulted in people's fear of visiting certain forests.

Also in other countries have criminals used city forests and nature as hideouts. Hampstead Heath, a common comprising moors, open lands and woods near London, was associated with gangs of 'highwaymen' from the 17th until the 19th century (A history of, 1989). At the beginning of the 20th century, the *New York Times* ran a story about an alleged assault on two women in the Bois de Vincennes of Paris (A queer crime, 1901). The assault took place in an area called the Labyrinth, a wooded and secluded area that had been popular as "sheltered rendezvous for the abominable wretches [read: prostitutes, Cecil Konijnendijk] who haunt the wood". This case added to the long list of dangers associated with the Bois de Vincennes as well as with the Bois de Boulogne, according to the newspaper: "There is no doubt that the two woods are places of great peril after dark."

In his history of Phoenix Park in Dublin, a large city park which harbours several smaller woods, Nolan (2006) dedicates an entire chapter to crime. However, before embarking on some rather grizzly stories, the author stresses that crime is bound to happen at places where many people gather. Thus, the urban park or city forest, as part of urban life, will be host to crime just like other parts of the city. Nolan writes that in former times, Phoenix Park was deemed so unsafe because of 'highwaymen' (robbers) that people required escorts when passing through. An especially dangerous patch was so-called Butchers' Wood, a famous hideout of robbers. The name Butchers' Wood did not derive from murderous activity, however. The Wood was named after the mayhem caused when butchers from the city markets gathered amongst the trees to settle disputes. These disputes were fought out in ritualised battles with tools of the butchers' trade, according to sources. In May 1882, Phoenix Park was the scene of a famous political murder in a period of revolutionary violence. The Chief Secretary and Under-Secretary, two of the highest ranking representatives of British rule, were both stabbed to death by revolutionaries. During the 20th century, two murders in Phoenix Park made the headlines. One was the murder of a young nurse who was sunbathing in the park by one Malcolm McArthur. The other was the violent assault on two German backpackers in 1991 by a gang of thieves, resulting in the death of one of the victims.

While murderers, robbers and other criminals were obviously seen as unwanted elements of city forests, there is another side to their story. The freedom offered by city forests can lead to crime, but it can also enable activities that may be illegal, but find support amongst at least part of society. The bands of thieves and robbers who challenged unpopular medieval rulers and their henchmen, for example, were sometimes seen in a more positive light by peasants who had lost their traditional hunting and common rights to the lords. Classic examples of these heralded outlaws, mythical or real, were Dick Turpin, who had Epping Forest near London as one of his hangouts, and Robin Hood of Sherwood Forest (e.g. Schama, 1995; Green, 1996; Muir, 2005). The Waltham Blacks were a gang of deer poachers making Epping Forest unsafe (Muir, 2005). In the tales about these legendary figures, forest rangers and wardens often feature as 'bad guys', representing the greedy landlords (Jeanrenaud, 2001).

In some cases, however, foresters did play a more positive role in folklore and fairy tale. One can think of the story of Little Red Ridinghood, where the forester comes to save the heroine in the end. A modern 'forester myth' was born a few years ago in the Walloon region of Belgium. The convicted pedophile and murderer Marc Dutroux, hated by the entire country, managed to escape from custody. He was captured in the middle of an extensive forest region by local forester Stéphane Michaux (Gazet van Antwerpen, 2007).

On a more artistic note, showing that crime can provide inspiration, the British rock band Genesis based their song "The Battle of Epping Forest", part of the 1973 album 'Selling England by the pound', on an actual London East End gang fight that took place in Epping Forest (The Battle of Epping Forest, 2007).

According to Lehmann (1999), it is typical for forests especially in times of war, crisis and major changes as well as of economic troubles, to turn into areas of resistance and withdrawal for semi-legal and illegal activities. Immediately after World War II, for example, forests around major German cities were made unsafe by gangs of youths and unemployed. According to Lehmann, 'border forests' ('Grenzwälder') arose after the setting up of the Iron Curtain. These forests were not only situated on the border between countries, they were also placed in the margins of society and attracted different kinds of illegal activities. During wars, forests have often become sites of atrocities, acting as settings for mass executions and mass graves hidden from the public eye (Jones & Cloke, 2002). In 1940, for example, up to 26,000 Polish officers, cadets and other officials were executed and buried by Soviet troops in the Katyn Forest near the city of Smolensk (Coatney, 1993).

As already shown for the case of Dublin's Phoenix Park, political tensions can also evoke fear of using city forests and parks, especially if they result in violence. The Jerusalem Peace Forest in Israel was created during the 1970s to commemorate the conquest of the old city by Israeli troops during the Six Days War. The forest used to be popular with all parts of the population and all religions, in spite of obvious tensions in society. A *Newsweek* journalist wrote that "the beauty of the setting seems to dispel all possibilities of violence". However, this changed dramatically in February 2002, when a Jewish student was murdered by Arab youths. This led to restrictions on forest use, the presence of armed guards and a major decrease of recreational use (Hammer, 2002).

The 'criminal side' of urban woodland can hamper its recreational use. According to work in Britain (reviewed by Jorgensen & Anthopoulou, 2007), urban woodland is perceived as fraught with dangers including intimidation from groups of young people, and physical or sexual assault. Thus some people are deterred from going to the woods on their own. Fears associated with possible assaults are compounded by the idea that if anything would were to happen, no-one would come to people's aid. Such a picture emerges from a study in The Netherlands. Visitors of the Haarlemmerhout city forest in Haarlem were asked how safe they felt in the forest. Half of the respondents answered that they felt unsafe or not really safe in the most densely vegetated part of the forest (Nibbering & Van Geel, 1993). When asked about their main fears, most people said they felt threatened by other visitors.

Different forest users have differing types of levels of fears. Especially women are often afraid to go into forests by themselves, as they fear for their personal safety and are especially afraid of sexual assault (e.g. Jorgensen et al., 2006; Berglund, 2007). In a project by Burgess (1995), thirteen women's groups from different cultures went on guided forest walks in woodlands near London and Nothingham. The groups included Asian and Afro-Caribbean women. At the start of the project, it became clear that women of all groups feared to be in woodlands on their own, not in the least because of crime and safety concerns. The sense of enclosure typical for woodland was seen as both good (as it stimulated tranquillity and the feeling of being away) and bad (as women could feel trapped or threatened). Women from the ethnic groups needed to be in much larger parties before feeling safe than the other women who participated in the study.

Interestingly, a study by Jorgensen and Anthopoulou (2007) indicated that elderly people (over 65 years) in Britain do not seem to be more afraid of crime then other urban woodland users. This group does have some specific fears, but these are associated with old age, such as fears to stumble and fall (also Burgess, 1995).

Nature's Dangers

In 2006, inhabitants of the Dutch town of Eelde contacted local police after having encountered snakes of up to 2 m in length in their local forest and nature area. The snakes turned out to be Russian rat snakes that had escaped from a collector. These snakes were not dangerous, but obviously citizens were alarmed, especially because of the high number of snakes seen in nature. Information evenings had to be held, a network of 'snake catchers' was set up and staff from a local zoo were brought in to capture the snakes (Eelde zit met slangen, 2006).

Although most city forests in Europe harbour only a very limited number of natural threats, this story shows that the issue of natural dangers needs to be confronted. This is particularly the case in a time when 'wilder' – i.e. less managed – nature is brought back into our cities and towns (see Chapter 8 'The Wild Side of Town'). As also described by Van den Berg and Ter Heijne (2004), to many people nature poses a wide range of threats, from animals transmitting diseases to extreme weather. Negative experiences are related with (unexpected) meetings with animals, confrontations with natural forces, overwhelming situations – for example, related to the largeness of a forest – as well as with unpleasant situations in general.

Lehmann (1999) and Muir (2005) also mention people's fear of certain animals. It is understandable that medieval people feared the nearby forests that still harboured large predators. In 1493, for example, the city council of Hanover offered a reward of 25 shillings for each wolf that was killed in the local Eilenriede forest

(Fachbereich Umwelt und Stadtgrün, 2004). According to Muir (2005) the wolf has been the beast that inspired most dread, as also reflected in myths and fairy tales. Today, the expanding population of wolves in countries such as Sweden is still a cause of fear and conflict (e.g. Créton, 2006). In North America, the mountain lion is a much discussed predator, especially as urbanisation and sprawl have brought people closer to the habitats of this large cat. Cronon (1996b) tells the story of a mountain lion that attacked and killed a female jogger in the foothills of the Sierra Nevada in California. The complex relationship between urban people and predators is illustrated by the aftermath of this case, during which more money was raised for the mountain lion's cubs than for the children of the jogger. Although most city forests today will not be associated with large predators, the work of, for example, Burgess (1995) in Britain has shown that ethnic minorities and new immigrants sometimes associate city forests with the 'wild' forests harbouring dangerous animals in their country of origin. Debates in countries such as The Netherlands and Germany about the reintroduction of large predators show that many Europeans have similar fears.

Animals can also evoke fear as transmitters of diseases. Bats and foxes transfer diseases that pose important health hazards to the human population, such as rabies. Lyme disease (or borreliosis) and other tick-borne diseases are also increasing. At the end of the 1990s, between 40,000 and 80,000 Germans were infected each year with Lyme disease (Lehmann, 1999). Numbers of those infected have been increasing ever since. The increase in Lyme cases has led to widespread fear, sometimes leading to a decrease in forest use. Studies in North America have shown, for example, that 10% of the public in the state of Montana felt at risk of getting the disease, although the transmitting tick had in fact never been found in the State (Ransford, 1999).

Not only animals can be a health hazard. Urban vegetation can also negatively affect human health, as pollen from certain trees and plants can cause allergies. In Sweden, for example, an estimated one-third of the population has some form of (often pollen-related) allergy (Sörensen & Wembling, 1996).

The forests in many European countries were also set in a more negative light after the nuclear disaster at Chernobyl, which meant that people feared to go into forests because of pollution, and were strongly advised to refrain from eating berries or mushrooms. Lehmann's (1999) interviews with German forest users show that the fearful summer of 1986 was still very much on people's mind, at least at the end of the 1990s.

Another fear at the interface between the natural and the human is that of forest fires. Each year the media show frightful images of forest fires close to cities in the Mediterranean and places such as California and Australia. Obviously, where forests and other vegetation are close to where many people live, this can cause fire hazards during dry seasons. Moreover, the chances for careless users or arsonists are statistically higher the more people a forest attracts (e.g. Sieghardt et al., 2005). Wildfires have been increasing at the urban-wildland interfaces across the world, partly due to the expansion of peri-urban areas and tourist zones (Nowak et al., 2005; Vallejo, 2005).

From Fear to Excitement

Fear of animals and misbehaving people, as well as some more 'primeval' fears related to forests hamper the use of many city forests, at least by parts of society. Segments of the urban population refrain from using the local woods entirely or at least during parts of the day. How can these fears, some rational, some perhaps less rational, be overcome? How can city forestry help by managing risk and danger (see e.g. Bues & Triebel, 2004)?

Dealing with fears caused primarily by animals and other 'natural elements' are one area to look at. Ongoing efforts to reduce diseases transferred by animals, for example by controlling animal populations, is one area of action. A balance will have to be found between bringing the forest and 'wild' nature all the way to people's doorsteps and dealing with the associated risks of this, such as increases in tick and mosquito borne diseases, floods, fires and so forth. In areas such as North America, Australia and Southern Europe, fire management at the 'urban-wildland interface' has developed as a vast field of research of its own.

With regard to primeval fears of forests, of the dark and dangerous wood where one can get lost, several actions can be undertaken. First of all, awareness raising, education through guided walks and the like can help people become more familiar with the local woodland, find their way around and notice that their fears are perhaps not very rational. The previously mentioned work by Burgess (1995) involved guided walks with groups of ethnic women to stimulate the use of local woods. The women felt safer in a group and being accompanied by a local forester. They also said that they noticed that the forest does not have to be a dangerous place. Suggestions for improving forest use by all groups include improving signage and information, maps, community liaisons and policing of forests.

Milligan and Bingley (2007) studied the relationships between young adults and woodland, showing that many young adults see woodland as a scary place, creating anxiety and uncertainty. Young people have expressed to feel intimidated, trapped or shut in, especially in denser woodlands. Myths and fairy tales are among the factors enhancing fear for woodland. Obviously children are a key target group for reducing fears of forests, as they should learn from an early age onwards how to use the woods. In the Nordic countries and increasingly also elsewhere, forest schools are a common phenomenon to make children at ease in the forest. In these schools, children spent part of their biology and sometimes also other lessons in the forest (e.g. O'Brien & Murray, 2007). In some cases, kindergartens are permanently based in the forest and children are outside in the wood most of the day (see Chapter 10 'The Forest of Learning').

Actions can also be undertaken in terms of forest design and infrastructure. C.R. Jeffery coined the term 'crime prevention through environmental design' in the early 1970s (cited in Khurana, 2006). It is based on the idea that proper construction and layout of the environment can help inhibit criminal activity and improve community living. The first of three main principles applied is natural surveillance, which relates to increased visibility in green and other public spaces. The second principle is called natural access control and deals with the public's physical

Fig. 3.2 Vandalism and littering are common in many city forests, as here in the Lorettowald of Konstanz, Germany. Visitors get the impression that the forest is not managed or controlled, resulting in a negative spiral or neglect and crime (Photo by the author)

movements through a space. Finally, the third principle looks at territoriality and builds on the theory of 'defensible space' developed by Oscar Newman during the 1970s (Newman, 1972). Based on these principles, paths can be lighted, signs installed, clear orientation points added, ground cover and shrubs next to paths removed, and so forth, all to reduce fears of getting lost and of suddenly encountering animals or humans.

In study of crime and enforcement in parks and recreational settings in the United States and Canada, Pendleton and Thompson (2000) identified that there is a 'life cycle' of crime in green spaces. It all starts with a period where there is a threat of disorder and rise of fear. Typical for this situation are increases in graffiti, vandalism and littering. If green space managers do not take rapid action to improve order, there is a risk of escalation (Fig. 3.2). In the end, efforts to fight crime in green spaces will have to be larger-scale, as guardianship has to be assumed to 'reclaim' the park.

This reclaiming of the park or city forest is a process which involves mobilising the local community, as well as enhancing social and management control. Lack of control and lack of visible presence of management authorities is often mentioned as a major factor in fears of criminal behaviour (Fig. 3.3). A study in the Dutch city forest of the Alkmaarderhout indicated, for example, that visitors wanted to see the 'forest supervisor' (i.e. some sort of permanent presence of forest management)

Fig. 3.3 The visible presence of forest managers or law enforcement officers, as here in the case of a London park, helps reduce feelings of unsafety (Photo by Hannelore Goossens)

back in the forest (Alkmaarder Houte dupe, 1992). In the US, the national Forest Service lacks the human resources in terms of so-called law enforce officers to patrol its forests (Wing et al., 2006), which implies that other means of control need to be found, such as mobilising the local community. When visitors to the Highgate Wood of London see damages or vandalism in the wood, for example, they can report these online to the managers. Moreover, they can also track online what has happened with their report (City of London Corporation, 2007).

Many of these aspects were discussed at a recent conference on anti-social behaviour in British parks and green spaces (Smith, 2007b). The event focused on how to deal with behaviour and activities such as use of off-road motorcycling, vandalism, prostitution and drug taking. In order to 'reclaim the parks from the mindless', various speakers stressed that a partnership approach involving local communities, the police, park managers and others would be needed. Practical experience had shown that crime and fear of crime could be reduced through a series of measures. These can include improving maintenance standards to clear overgrown shrubberies, clearing out hiding places and providing a feeling of conspicuous care, and cleaning graffiti as soon as it appears. Other, more partnership-based measures are creating nature areas working with local primary schools, re-engaging the local community and bringing on board police, rangers, housing wardens, schools and parents to create a multi-agency approach and a strategy for park improvement.

However, there is a downside to many activities to prevent and reduce crime, as stated in a report on the social values of public spaces by Worpole and Knox (2007). People run the risk of being 'designed out' of the public realm. The authors state that the British government's emphasis on crime and safety in public spaces is depriving these spaces of their historic role as a place where differences of lifestyles and behaviour is tolerated. Moreover, certain groups of green space users – perhaps considered 'abusers' by some – are left out by definitions of 'communities' that exclude particular groups, such as young people or sex workers (see also Thompson et al., 2006). Approaches that strip public spaces of all features vulnerable to vandalism or misuse also actively discourage local distinctiveness and public amenity. In the specific case of city forests, care will have to be taken that the local city forest does not become overly 'controlled' and full of facilities, as this might take away from the 'natural' forest experience that many visitors prefer.

Local community involvement seems crucial in all approaches to managing danger. Studies of American cities have shown that crime rates are lower in places where settings regular share and display a strong proprietary sense of territoriality (Newman, 1972; Gallagher, 1993). Jones and Cloke (2002) show how a local community group 'reclaimed' the Arnos Vale Cemetery of Bristol, UK. The cemetery had gradually deteriorated, attracting unwanted 'uses' – e.g. three people pleaded guilt to taking a corpse from a vault – and vandalism. Female visitors to the cemetery, which had gradually come into use as outdoor recreation area, described parts of the area as spooky and sinister. A group called the Friends of Arnos Vale Cemetery was established and started to carry out surveillance and maintenance activities. This led to a reduction in criminal activity, although some legitimate users of the area started to feel 'excluded' by the group (see Chapter 11 'The Social Forest').

City forests are not only a setting for crime. Trees and (well maintained) green spaces have been found to sometimes contribute to a reduction in crime rates. Work by Kuo et al. (1998) in a Chicago housing estate, for example, indicated that higher densities of trees and better grass maintenance contributed to an increased sense of safety amongst residents.

Urban nature, by its very characteristics, should be allowed to keep its 'wild edge' and its role as adventurous space, offering exciting and less-controlled opportunities for play and recreation. As Chapter 6 'The Great Escape' and Chapter 8 'The Wild Side of Town' will show, city forests have always acted as space and as frontier in urban settlements, enabling feelings of freedom and activities that are not possible elsewhere in the city. At a conference on 'urban wildscapes' held in Britain, three central forms of paradox were mentioned as characterising these wildscapes: danger versus adventure, nature versus nurture, and junkyard versus paradise. It was stressed that doubt and uncertainty are an integral element of the experience of growing up for children and young people (Simson, 2007b). Studies indicate that wilderness and outdoor challenge programmes for youths help promote their self-esteem and sense of self (Jorgensen et al., 2006; Taylor & Kuo, 2006). Thus certain feelings of fear can be 'turned around' to add to the excitement of the city forest experience (Van den Berg & Ter Heijne, 2004). Visitors can be surprised by the unknown and find new courage and self confidence.

Today's entertainment society is full of exciting opportunities for people to enjoy themselves. The city forest will have to compete for users and attention on its own premises. Several recreation entrepreneurs and even forest management organisations have picked up this challenge, offering survival games, paint balling, role playing games, canopy walkways, and other exciting leisure activities. But the real adventure of city forests, which includes a more pleasant perspective on the Forest of Fear, remains to be fully developed.

Chapter 4
The Fruitful Forest

Forests, including city forests, have always provided our society with a wide range of benefits. Jones and Cloke (2002, p. 40) speak of "working trees" which offer a wide range of products and services. This book focuses on the many cultural and spiritual roles of forests. These can be categorised under city forest services, as in the case of forests providing opportunities for recreation. However, insight in the use of forest products is also very important for understanding the wider socio-cultural importance of city forests.

This chapter starts by describing the subsistence function that forests situated near towns have had for peasants and the urban poor until modern times. Next, efforts by aristocracy and others in power to appropriate city forests for their own uses, such as hunting, are discussed. The economic role of forest products in city development and national building is the topic of the third section. This is followed by an account of how municipal forestry, i.e. city forestry carried out by municipal authorities, developed, initially with economic considerations in focus. Finally, the current state of the Fruitful Forest as a provider of forest goods to citizens is discussed.

Subsistence Forests

As described in Chapter 1, medieval cities did not have many trees inside their walls. If trees were present, they were mostly to be found near temples, churches and in secluded private gardens, for example of monasteries. Most gardens had a clear production function, as they contained plants and trees producing fruit, flowers, medicines or herbs. All of these also carried symbolic meanings and cultural associations, often religious in nature (Lawrence, 2006).

The large class of peasants living near cities, as well as the urban poor, were dependent on nearby forests for fuelwood, fodder and grazing their cattle. Muir (2005), for example, describes how so-called 'hollins', holly groves or stands of holly (*Ilex aquifolium* L.) in larger woods were an important source of winter fodder. During the Middle Ages, the diet of the inhabitants of Gdansk included "the fruits of the forest" (Cieslak & Biernat, 1995, p. 25). Forests near cities thus had an important subsistence role (e.g. Buis, 1985; Westoby, 1989; Rackham, 2004). Forest use was regulated through customary rights (Jeanrenaud, 2001). Land use rights were complex, distinguishing between use of soil, timber, fuelwood, grazing rights, hunting rights, and the like. Often the land owner did not have all these different rights in his hands (Rackham, 2004).

Rackham (2004) and Muir (2005) describe the various types of wooded lands and their use during history. From the 'wildwood', which was not named, owned or managed, various forms of cultural woods emerged. As early as during the Middle Ages, many woods were managed intensively and conservatively. Most woods were transformed to coppice forests to produce the desired products, for example, to support farming. As explained in Chapter 1, 'Forest' was initially used as legal term for the wood pastures and hunting areas of the King. Forest could be wooded, but this was not necessarily the case. User rights were more diverse in the case of common lands, which were used by commoners ranging from landless squatters to rich farmers. Commons included 'wooded commons', where the same tract of land was used for growing trees and for grazing (Rackham, 2004). Epping Forest near London, where many people grazed their animals, was an example of an area with common rights, although it initially was a 'forest', i.e. a royal hunting area (Fig. 4.1). A forest law instated by King Henry I (around 1130) led to a conflict, as restrictions were placed on forest use by peasants. However, rights of commoning were made clear and so-called 'verderers' administered Epping for the King, whilst also protecting the rights of peasants (Green, 1996). For small communities during medieval times,

Fig. 4.1 Epping Forest near London is an example of a city forest that started off as a wooded common. Large pollarded trees such as the ancient beech tree shown here provided commoners with a range of wood products for construction, heating and other purposes (Photo by Alan Simson)

the possession of local woodland was of great importance. Common rights to wood and timber were jealously guarded (Wiggins, 1986).

As the majority of the population lived outside cities, rural forests were more important from this subsistence point of view than city forests in a true sense. Exceptions of forests in near towns included the Eilenriede forest of Hanover, Germany, which was used very much as a wooded common for extracting wood and for grazing purposes (Hennebo, 1979). The 'masting' of pigs, also known as pannage, was an important use of many forests, as reflected in medieval laws (Schama, 1995; Rackham, 2004; Baeté, 2006). Masting helped remove the green and poisonous acorns from woods that were also used for grazing cattle and deer (Muir, 2005). Pig masting was also very important in the mentioned Eilenriede of Hanover, where oaks were 'set free' to improve mast (Wolschke-Bulmahn & Küster, 2006).

Cities expanded and became more dependent on nearby forests. For example, the founding of the German 'twin cities' of Berlin and Cölln in the 13th century led to a growing demand for construction and firewood, litter and resins (Cornelius, 1995). Forest areas such as the Grunewald were under considerable use pressure as early as during the 14th century. Various communities and members of the aristocracy held grazing rights, implying that 111 horses, 624 cattle and no less than 4,000 sheep could be grazed in the relatively small forest.

Forests were also important for the emergence of industries. Across Europe, small-scale woodland societies developed that burned charcoal which fueled primitive fireworks, stripped bark used for tanning, drew fuel for glassworks and breweries and felled timber beams and supports for city houses (e.g. Rackham, 2004). Although these local communities resided in the forest, they had close ties to nearby towns, as these were their primary customers (Schama, 1995). The rise of the metallurgic industry added to the demands for fuelwood and thus placed more pressure on forests (Jeanrenaud, 2001). The forests of Baden-Wuerttemberg, Germany, which include what are now the city forests of Freiburg, hosted such small-scale industries, including family-based glass industry. But until the 19th century, the forests also continued to offer nutrition to the people. Grazing proceeded as well, as did the extraction of litter, resin, tar and fuelwood (Ministerium für Ländlichen Raum Baden-Württemberg, 1990). Products that were taken from Epping Forest by peasants and the urban poor included wood for household carpentry and fuel, as well as beech nuts, fungi, birch sap and elderberries for wine. Grazing and pollarding of trees were common activities (Rackham, 2004).

Some of the subsistence uses of nearby woodland were regulated by means of early municipal ownership of city forests. Hanover was one of the few cities in what is now Germany that owned forests as early as in 1371. The Eilenriede forest, for example, was used for producing wood and as area for grazing. A system of fences, moats and guarding towers was set up to help control and protect the Eilenriede from illegal use and the mixing of cattle and deer (Hennebo, 1979; Wolschke-Bulmahn & Küster, 2006). The Swiss city of Zurich also made early attempts to extend its control over nearby forests such as the Sihlwald, as many of the small industries in the city required a steady supply of wood (Irniger, 1991).

Appropriation of City Forest Use

According to Lohrman (1979, cited by Irniger, 1991), there was no shortage of wood in Western Europe until about the 12th century. However, with further population growth, the demand for wood for industry, construction and fuel grew. Additionally, as more mouths needed to be fed, forests also fell victim to agricultural expansion. These developments meant that rulers and the aristocracy started to worry about forest overexploitation. They were especially concerned with maintaining forests as hunting domains, as well as with keeping them as timber reserves. Initial protective measures and regulations were installed and several hunting forests and hunting parks were (partially) closed off from the general public (Jeanrenaud, 2001). As mentioned in Chapter 1, 'forests' were (mostly larger) areas where hunting rights were reserved for the kings. In the case of hunting parks, rights were licensed by the king to the aristocracy (Muir, 2005). In both cases, hunting areas were more stringently protected from poaching and other 'abuse'. Great attention was paid to boundaries and security and many manorial court records are full of (often petty) trespasses against royal or aristocratic woodland (Rackham, 2004). Three types of trespassing against forestal rights were distinguished in England, namely trespass "against the venison" (against the deer), "against the vert" (damaging of deer habitat) and "assart", i.e. appropriating part of the physical forest for private use (Rackham, 2004, p. 171).

To an increasing extent, even the wooded commons were appropriated by rulers and the upper classes, leading to conflicts with local communities of peasants and poor. This issue is discussed in greater detail in the chapter 'The Forest of Power'. Rights of use became formalised. When forests became smaller and situated nearer to urban centres and dense populations, their value increased (Bitterlich, 1967). In many cases, only those forests protected by the elite would survive the growing pressures of city expansion, as in the case of the royal woodlands of Prague (Profous & Rowntree, 1993; Kupka, 2005).

Use of forests in the area that is now the Zoniënwoud south of Brussels became regulated through a system of heavy penalties during the 13th century. A special register was kept to assist the protection of trees and the control of grazing (Van Kerckhove & Zwaenepoel, 1994; van der Ben, 2000). Elsewhere, the ruling Danish king limited free use of the forests near Tallinn as early as in 1297 (Meikar & Sander, 2000).

As mentioned, wooded areas near cities and towns, where kings and rulers had their courts, were popular hunting domains. This was an important factor in the protection of city forests. Gdansk in Poland was the residency of the Pomeranian princes and thus these princes with their retainers and attendants played the foremost part in the big game hunt. Game for hunting in the forests near Gdansk initially included, elks, roe deer, wild boars, bears, beavers and hares, with wolves and lynxes as special trophies (Cieslak & Biernat, 1995). Like the Pomeranian princes, King Henry VIII of England was an enthusiastic hunter. In 1543, he had the Queen Elisabeth Lodge built in Epping Forest (Slabbers et al., 1993). Other popular royal hunting areas of those days included the Forêt de Saint-Germain near Paris, which was used by the French kings (Slabbers et al., 1993). In what is now Belgium, special 'wild boar parks' were established for hunting purposes. These needed to be enclosed because of damage to agricultural lands caused by roaming boar (Baeté, 2006).

Pressures on city forests increase further after the Middle Ages. Royal forest owners continued to close off their properties and to chase 'illegal' users away (Hennebo, 1979). Walls were built around forested estates such as the Forêt de Saint-Germain (Slabbers et al., 1993) and the Bois de Vincennes (Derex, 1997b) near Paris, France, and Belvoir Park near Belfast, Northern Ireland (Simon, 2005). Jægersborg Dyrehave near Copenhagen was fenced and practically closed to the public during the reign of King Christian V (Møller, 1990). During the political unrest of the 17th century, the English royal parks were given free, but part of them remained in the hands of speculators. After the Restoration, the parks came in royal hands again. Some activities in them remained reserved to the upper class. King Charles I, for example, fenced the area of Kensington Gardens and put game in it (Hennebo & Schmidt, s.a.).

Elite hunting remained the main use of many city forests until more recent times. Under the 'Soldier King' Frederick William I (1713–1740) hunting continued to be important in the area surrounding Berlin (Cornelius, 1995). The function of Berlin's Tiergarten was still that of a near-city 'Wald-und Jagdrevier' (forest and hunting district) (Nehring, 1979). According Lieckfeld (2006), the so-called 'Hegejagd', aimed at trophy hunting, remained a rather elitist activity in Germany and elsewhere until far into the 20th century. The Schorfheide area close to Berlin, for example, was Marshal Hermann Goering's favourite and exclusive hunting area during the Nazi-era. The woodland of Bos ter Rijst in Flanders, Belgium, harbours an enclosed wild boar hunting domain which came into the hands of a developer from Brussels in 1973. His hunting parties with associates were allegedly well known (Baeté, 2006).

However, hunting was not the only city forest use appropriated by the upper classes. By the end of the Middle Ages, timber production had become an important function of some city forests. The value of wood had rapidly increased, primarily due to the increasing economic demands of cities during the 14th and 15th century (Irniger, 1991). Timber exports were important, for example, for the city of Gdansk (Cieslak & Biernat, 1995), while in mid-16th century Prague, acceleration of housing and mining construction created a high demand for wood (Profous & Rowntree, 1993). Timber cutting and wood trade in Berlin reached a high when Frederick the Great run into major financial troubles after the Seven Years War (Cornelius, 1995). City forests also catered for other demands for products. Park Valcberg (now Valkenberg), a forest estate in the Dutch city of Breda, was mainly used by the aristocracy for hunting and pleasure, but it also produced vegetables and fruits (Dragt, 1996).

With the rise of a class of merchants and industrialists, a new group of actors claimed their stakes. Gradually, the bourgeoisie and private entrepreneurs started to purchases city forests, for recreation and prestige reasons, but also for wood production purposes. During the Industrialisation, coal companies in, for example,

England often owned woodland, primarily for producing support beams for the mines (Wiggins, 1986). Forestry had become of further economic and scientific interest during the 19th century based on the concept of sustained yield. In many countries forests were enlarged for industrial purposes (e.g. Jeanrenaud, 2001).

During the late 19th century, major land owners started buying and developing wooded estates near cities as investments and for land speculation. In the case of the Dutch city of Arnhem, for example, a handful of influential families patiently bought and assembled areas to create large estates (Schulte & Schulte-van Wersch, 2006). Later these would be bought by the city and become popular city forests, such as Sonsbeek.

City Forests and Societal Development

In his account of the importance of timber for civilisation over time, Perlin (1989) mentions that trees have been the principal fuel and building material for almost every society for over 5,000 years. City forests, in particular, played an important role in the building of first city states and later nation states, as they provided a nearby source of timber and fuelwood. Timber was needed for house building and to some extent the construction of ships (Rackham, 2004). Wood was also a crucial source of heating and energy. The emergence of an iron-industry based on charcoal has already been mentioned.

Wood was invaluable during the times of the first city states. Mesopotamian cities such as Uruk and Ur exploited the wood from local forests (Perlin, 1989). Wood was often in such high demand during periods of expansion that its value approached that of precious metals and stones. A pattern emerged of city development being based on nearby forest resources. The original forest growth near Knossos on Crete became replaced by a secondary forest of cypress (*Cupressus* spp.). The city states of Ancient Greece had a large demand for timber to build ships, which led to overexploitation of nearby forests. After the defeat of Athens in the Peloponnesian War, the mountains near it were depleted of forests, leading Perlin (1989, p. 93) to write: "[w]here once wolves roamed, hunters could now not even find a single rabbit to spear". Deforestation led to erosion and detrimental effects on agriculture. This also resulted in a new ecological consciousness and to new laws and regulations. The latter were particularly needed for the remaining sacred groves (see Chapter 2 'The Spiritual Forest'). For the nearby populace trees in these groves were a last resort when timber and fuelwood had run out.

Rome and its surroundings were once covered by woods as well. Forests provided Rome with the material essential to its growth. But during its expansion, houses and buildings began to cover the hills where trees once stood. Extensive ranging and agriculture started to replace woods. Still the demand for timber was bigger than ever, not only for construction, but also for baths and the glass industry. Influential statesmen such as Cicero expressed their concern over the decline of

Rome's woodlands in Senate debates (Perlin, 1989). Pliny introduced principles of silviculture, including cultivation of the willow which would go a long way in meeting the needs of farmers. As a result, the value of trees in Rome "rose in the opinion of both philosophers and thieves" (Perlin, 1989, p. 120).

The mistakes of the past were repeated by medieval towns in Europe, in spite of efforts to conserve forest (and thus wood) resources (e.g. Corvol, 1991). The Venetian Republic, for example, was heavily dependent on wood for its fleet (Perlin, 1989). Regulations were imposed on nearby forests and exportation of wood cut from the foothills and mountains directly north of the city became prohibited. Still, the regulations were only partly successful due to continued cutting and burning (to clear land for agriculture) by local communities. Deforestation continued, also in more remote forest areas in what is now Slovenia. One of the consequences of deforestation was the silting of the lagoon, the city's natural harbour.

England provides yet another example of increasing exploitation of forests by rapidly developing towns. Woodland decline, primarily due to population growth and the associated rise of agriculture, had come to a hold in 1349, when the Black Death (the plague) hit Britain (Rackham, 2004). Later, however, the population grew again. The large combined demands on forests by Elizabethan society related to building of houses, furniture, casks, carts, wagons, coaches, making iron, burning of brick and tile (Perlin, 1989), and also navigation, although Rackham (2004) considers the demands of shipbuilding of relatively less importance. In 1576, the mayors of Hastings, Winchelsea and Rye sent a letter to the local Lord, expressing concern over the building of new ironworks. These would compete with local timber cutting. The mayors even took their complaints to Parliament. The London fuel crisis of 1643 led Parliament to appoint officers to oversee the felling of trees. It also made available to Londoners the woods formerly belonging to the King, who had been temporarily removed from power. Parliament opened up a 60-mile ring around London as 'fair game' for felling wood for fuel. Concerns about timber supply led to people such as John Evelyn calling for better management of forests and forest establishment. The Great Fire of London of 1666 tested wood supplies even further (Perlin, 1989). However, the main threat to wooded areas during those times, and especially those in and near cities, was conversion to agricultural land, so that more mouths could be fed (Rackham, 2004).

As mentioned earlier, the 18th and especially 19th century brought a further increase in the importance of wood production. Cities such as Berne, Switzerland, acted as a 'black hole', "devouring enormous amounts of energy and raw material from their surroundings" (Jeanrenaud, 2001, p. 19). Berne's 12,000 inhabitants consumed about 50,000 m^3 of wood annually by the year 1800 (Köchli, 1997, cited by Jeanrenaud, 2001). Eighteenth-century Basel consumed even more wood, as its demand for fuelwood was estimated to be 120,000 m^3/year (Haas, 1797, cited by Meier, 2007). During the same era, Gdansk was one of the major centres for processing and exporting woods. Forests near the city disappeared, for example because of the timber industry, but also because of extension of the port. The timber industry of Gdansk largely disappeared during the second half of the 19th century, not in the least because of lack of supply (Cieslak & Biernat, 1995; Szramka, 1995).

City forests outside Europe also played an important role as local source of timber for construction and heating. Houle (1987) describes the use of the forests around the emerging city of Portland, Oregon, United States, in the area that would later become the city forest of Forest Park. Settlers used wood from the forest for building and fuelwood, for example for steamboats crossing the river.

In line with this growing pressure on forests, nations and cities commenced to recognise the need for conservation and for applying principles of sustained yield. At the beginning of the Industrialisation, more forests were brought under public control, leading to a diminishing of individual rights and to communal use of forests (Hennebo, 1979). Many 'badlands' were afforested under the growing influence of the State (Buis, 1985).

Forestry professionals that followed the principles of sustained yield and rational forestry entered the stage. In Berlin, for example, a 'forstliche Finanzverwaltung' (forest finance authority) was set up as early as in 1723, as described by Cornelius (1995). It operated from the principle that the income from the domains owned by the Prussian Kings had to be increased. This was impossible with traditional wood trade. But the development of the shipbuilding industry held promise for the development of sawnwood markets. A considerable amount of wood was exported from Germany to countries such as Holland. Forests in the surroundings of the rapid expanding residency of Berlin were used for wood extraction, as well as a source of energy. Prices for fuelwood increased and a monopoly on it was imposed to safeguard the position of the poor. Wood from the Berlin forests was also used for the production of glass – for which potassium was needed – and iron. The Brandenburger Kurfürsten and Prussian Kings tried to create a domestic basis for iron production, in spite of the often bad quality which was produced. Tar production was also practised. Minerals were not only extracted from the forests in the form of wood, but also by increased litter removal for summer stable feeding. Litter extraction was particularly detrimental because of Berlin's poor soils.

The Rise of Municipal Forestry

As shown, growing concerns over securing a continued supply of wood led to an increase in the interference of cities in forest management. As forest ownership offers arguably the best means of controlling what happens with the forest, this increased interference also gave rise to more municipal forest ownership. As discussed earlier, communal forest ownership in Europe has a long history, dating back to the times of the wooded commons (Jeanrenaud, 2001). In some cases, forests became municipal ownership through gifts, purchases, war adjustments, and after dismantling of church institutions during the Reformation (Hosmer, 1988; Konijnendijk, 1999).

It is easy to understand why city authorities choose to own and manage forests. Settlements had to satisfy their own demands for construction and fuelwood, at least until transportation was improved by means of the railroads. In his model of land use, Johann Heinrich von Thünen famously situated a ring of production forests close to the city centre, demonstrating the importance of short transportation routes for timber (e.g. Bues & Triebel, 2004). Lloret and Mari (2001) describe how the Catalan city of Tortosa, Spain, controlled and managed the pine forests on the nearby mountain range as early as during the 14th century, even though these forests were owned by the Crown. Forest management was conservative and strictly regulated and there was a strong fire suppression policy, clearly indicating that wood obtained from the forests was seen as more important than agriculture and grazing.

Already during the second half of the Middle Ages, quite many cities and towns in what is now Germany and The Netherlands owned forests (Holscher, 1973; Hennebo; 1979, Alkmaarder Houte dupe, 1992; Meikar & Sander, 2000). These included city forests such as the Alkmaarderhout (Alkmaar), the Haarlemmerhout (Haarlem) and the Eilenriede (Hanover). In the case of the latter, feudal rulers transferred the Eilenriede forest to the city of Hanover in 1371. In the so-called 'Schenkungsurkunde', it was stated that the city had the right, or rather the obligation, to expand the area of the forest (Fachbereich Umwelt und Stadtgrün, 2004). Also in Germany, Erfurt has owned forests since 1359, while Nuremberg purchased a public meadow in 1434 and planted it with lime trees (Konijnendijk, 1999; Forrest & Konijnendijk, 2005). The city council of Freiburg started controlling the use and management of forests surrounding the city at the end of the Middle Ages (Brandl, 1985). Examples of early municipal forest ownership outside Germany and The Netherlands include the Slovenian city of Celje, which received the rights to nearby forests in 1451 (R. Hostnik, 2007, personal communication). Another example of long-standing municipal forestry is the Sihlwald near Zurich, Switzerland, as the first 'Zurich Forest Law' dates back to 1483 (Irniger, 1991).

The rise of municipal forestry led to changes in the prioritisation of forest functions and thus to changing forest management and structure. This is illustrated in the case of the mentioned Sihlwald by Irniger (1991). Already during the second half of the Middle Ages, more 'city oriented' criteria had started to determine forest use and management. Traditionally favoured forest uses such as masting and grazing now came secondary to supplying the city with wood. Zurich's city council claimed its interest in the forest. It drew up the first regulations and installed forest keepers. In 1424, for example, the first 'Sihlher' (Sihl master) was elected by the city council and a group of influential citizens. This Sihlher was responsible for ensuring a steady wood supply (and most notably fuelwood) from the forest to the city. He also controlled the cutting of trees and the floating of timber on the Sihl river.

Growing interference of cities inevitably led to conflicts between them and neighbouring, rural communities, as well as between cities and the religious institutions that still owned many forests up until the Reformation (see Chapter 2 'The Spiritual Forest'). Forests in and near cities elsewhere also changed form due to the rise of city forestry. In France, for example, peri-urban forests are among oldest cultivated and managed forests (INRA, 1979). In Germany, the Saarkohlenwald of Saarbruecken was shaped by traditional forestry uses, as well as by small-scale mining activities (Kowarik, 2005).

Cities and other municipalities sometimes managed to turn city forestry into a profitable activity. The Swiss city of Winterthur has a long history of city forest ownership and management. Its forests have supplied high-quality timber for manufacturing (Hosmer, 1988). Some German villages and towns have used the annual net proceeds from forests to pay for public expenses such as schools, roads, and the like (Hosmer, 1988).

By the time when municipal forest ownership became more common, i.e. during the 18th and particularly the 19th century, many city forests were in a deplorable state because of wars, over-use and lack of management (Hennebo, 1979). This was in line with the overall condition of woodlands across large parts of Europe, where for example coppicing had fallen into decline due to loss of markets for its products. Some surviving local woods, however, still catered for local timber and wood needs (Rackham, 2004). The city council of Hamburg, Germany, imposed various regulations at the beginning of the 18th century to stop overcutting and forest degradation. Corruption was one of the reasons that overcutting had continued to take place. As forest loss continued also after the acceptance of the first forest regulations, the city made further efforts, such as setting up a forestry administration and reforestation with fast-growing conifers. These efforts had success and during the Industrial Revolution, the Hamburg forests played a crucial role as source of energy and raw materials. This growing importance, in its turn, helped to further increase the forest area (Walden, 2002).

Over time, of course, the priorities for municipal forestry changed. Urbanisation led to new demands in terms of leisure activities, whilst the Industrialisation meant that fuels other than wood came into use. Holscher (1973) describes how 'community forests' in Europe evolved into places to stroll or hike, ski or participate in other forms of recreation. Historically, many Europeans, particularly the landless industrial workers of the larger cities, had few places to go for recreation and relaxation. Municipal forests ringing the cities were the only available sites that offered escape from the urban environment. As European cities were often compact, the surrounding woods were easily reached. Heske (1938) mentions that, by the first half of the 20th century, city forestry prioritised the 'spiritual and bodily hygiene' of the urban population rather than obtaining a financial return. An example of this is the involvement of the City of London Corporation in Epping Forest. In 1878, its Epping Forest Act ensured Epping Forest's role as place of public recreation, thus acknowledging and extending its customary recreational use (Layton, 1985; Green, 1996). Much more about recreational uses is told in Chapter 6 'The Great Escape' and Chapter 11 'The Social Forest'.

Other forest uses, and in particular forest services, have gradually also come into focus, such as the provision of drinking water and the protection of nature. At the beginning of the 20th century, for example, the city of Munich bought farmland located about 30 km outside the city with the aim to secure its drinking water supply (Lieckfeld, 2006). The area was afforested and turned into the 'Münchener Stadtwald' (Munich City Forest). Vienna's 'Quellschutzwälder' (Waterhead Protection Forests) are situated at an even greater distance from the city proper (Ballik, 1993). Forest establishment and management have also been used as employment projects in times of economic crisis. This was the case for the Amsterdamse Bos, which was created as a large-scale labour-provision project during the Depression (Balk, 1979). Cities have also turned to their forests and other green spaces to attract development, rich tax payers and businesses.

One special role that city forests have had over time is that of providing a resource to fall back on in times of crisis. There are many examples from the distant and not-too-distant past of how cities were dependent on their forests for timber, fuelwood and even dietary supplements. The forests near Athens, for example, lost their tree cover during the many wars (Strzygowski, 1967). During and immediately after World War II, large amounts of timber and fuelwood were taken from the forests near many cities. Partly this was done by their inhabitants (for fuelwood), and partly as part of the war effort (timber, cutting for making space for war installations). The forests near Munich and Berlin were largely cut for these purposes (Cornelius, 1995; Lieckfeld, 2006). Parts of the Amsterdamse Bos nearest to the city of Amsterdam were cut during the so-called 'Hunger Winter' at the end of World War II (Balk, 1979). One of World War II's most famous sieges was that of Leningrad (now St. Petersburg). Kitaev (2006) describes how this siege from 1941 to 1945 led to the loss of about 100,000 trees and 400 ha of green space.

More recent is the case of Sarajevo, Bosnia-Hercegovina where thousands of trees were cut for fuelwood during the long siege by the Bosnian-Serbian army

Fig. 4.2 During the long siege of Sarajevo, Bosnia-Hercegovina by the Bosnian-Serbian Army, the city's inhabitants were dependent on local trees and woods for heating. Because of snipers and mines, however, collecting fuelwood was a very dangerous undertaking (Photo by the author)

during the 1990s (e.g. Cordall, 1998; Fig. 4.2). During the third winter, the city had run out of electricity and gas. A newspaper article (Alleen tussen de frontlinies, 1994) tells the story of a young soldier collecting wood in the no man's land around the city, taking the risk of being shot by snipers. The Horsh Beirut Forest in the centre of Beirut, Lebanon has suffered from many wars and conflicts during the past decades. Bombs and overcutting for fuelwood have had their negative effects, although the forest has been replanted every time (Bakri, 2005).

War and post-war city forest use has also included supplementing diets. Many forests near German and Austrian cities, for example, were turned into food growing gardens in the years following World War II (e.g. Rotenberg, 1995). Lehmann (1999) describes how immediately after that war, German city dwellers turned to city forests to supplement their diets by picking mushrooms and berries. This influenced an entire generation's link with the forest. The author notes that mushroom picking was initially not practised in Germany. The custom was introduced after the war, with refugees from German-speaking areas in other countries and Eastern European immigrants.

In post-war times, overcutting of forests often continued. In Germany, fuelwood from forests was distributed to farmers and refugees after World War II (Lieckfeld, 2006). Also during those years, many German and Austrian city forests continued to be overcut as part of compensation fellings (for settling war debts) by allied forces (e.g. Lehmann, 1999; Lieckfeld, 2006). Allied occupation had various impacts on city forests. An interesting case is that woods near Russian army camps tended to have lower game pressures due to hunting, thus allowing so-called 'Russian oaks' to regenerate (Lieckfeld, 2006). The relicts of war were also affecting those fighting the heavy fires threatening the suburbs and city of Dubrovnik, Croatia, during the summer of 2007. Fire fighting efforts were hampered by remaining land mines from the Balkan wars of the 1990s (Dubrovnik menaced by, 2007).

Crises other than war have resulted in overexploitation of forests. In Armenia, an energy crisis led to overuse of urban and peri-urban beech and oak forests for fuelwood. Due to a shortage of gasoline, the transportation system collapsed, which hampered the normal supply of fuelwood. Today, Armenian cities are surrounded by heavily exploited, degraded forests. Even roadside trees, including poplars, have been cut (Davis, 2007).

The Fruitful Forest Today

As this chapter has illustrated, city forestry has its roots in the efforts of cities to secure the supply of forest products, most notably timber and fuelwood. In the planning and management of city forests today, focus is very much on the social and environmental services that city forests provide. Recreation, water protection and landscape aesthetics are among the principal goals of modern city forestry (e.g. Konijnendijk, 1999). In some cases, this new focus has led to a backlash against the production roles of city forests. The increased dominance of urban ideas, values and lifestyles has led to resistance against rural production processes, including those of forestry (Paris, 1972; Otto, 1998). The clash between urban and rural values and ideas also emerged from a case presented by Jeanrenaud (2001). In 19th century Switzerland, foresters were initially seen as 'enemies' by the rural population, as they were regarded to represent urban interests.

However, in spite of the rightful emphasis of city forestry on providing social and environmental services, there is still a role for the Fruitful Forest as well. A survey of Finnish municipalities, for example, showed that more than half of all municipal forests are still managed commercially, with emphasis on timber production (Löfström et al., 2006). In the Forêt de Saint-Germain near Paris, wood production also still plays an important role. The managing French State Forest Service generates important income from timber, supplemented by income from fishing rights, land rents and a golf course (Slabbers et al., 1993).

Production in modern city forestry can be highly innovative. The British town of Milton Keynes produces its own, high-quality cricket bats (Konijnendijk, 1998). Also in Britain, the Jaguar car company sponsored the establishment of the Jaguar Lount Wood, a community forest near the English town of Lount with walnut trees (*Juglans* spp.). The company thus made a connection with the wood used inside their cars (Forestry Commission, 2007).

One innovation links back to the history of city forests as sources of energy. In a time when renewable energy is being promoted, forest biomass has become recognised once again as potential source of bio-energy. This is the case for the forests of Basel, Switzerland, where a new woodfuel plant for heat generation was to start its operations in 2008. The Basel forest service already sees energy wood as a key component of the city's forest policy (Meier, 2007). In the German Ruhr area, part of the new woodlands that are being established will be energy woods, producing biomass for heating and electricity (Lohrberg, 2007). In the Croydon district of London, 'tree waste' from arboriculture and woodland management is turned into woodfuel at a community tree station. The station is capable of generating up to 10,000 t/year of high quality woodchip from the waste material (Jeanrenaud, 2001; DEFRA, 2007). Also in the case of England, Simson (2001b) describes the setting up of a biofuel plant in Yorkshire, England, that runs on 75% woodchips. Initiatives like this reconnect city forests with the local economy. These developments call for new approaches in the design of productive landscapes that combine, for example, growing biomass with other forest functions (Rook, 2007). Various European cities have had their forest management and forest products certified by organisations such as the Forest Stewardship Council (Konijnendijk, 1999).

Hunting is one of the traditional uses of city forests. In fact, as this chapter has described, it has played a key role in the development and protection of city forest landscapes. The importance of hunting from the perspective of changing power relations is the topic of Chapter 5 'The Forest of Power'. Today, hunting is a far less dominant use of city forests, although it still does play a role. A highly urbanised country such as the United Kingdom currently has 800,000 registered hunters (FACE, 2007). Another urban society, that of Denmark, has 180,000 hunters, although its total population is less than 5.5 million (Jensen et al., 1995). Many of these hunters live in urban areas. An older study in then West Germany, for example, found that 40% of all hunters were businessmen, managers and professionals with a higher income, mostly living or working in the city (Röbbel, 1967). When hunting takes place in city forests, it is usually carefully controlled. Hunting to control game populations in Fontainebleau, for example, is avoided during busy times and the special 'children's day' (Ruffier-Reynie, 1995). In Norway, each municipality has a 'game commission' which guides local hunting practices (Bosma & Gaasbeek, 1989).

Especially in Eastern and Northern Europe, the gathering of berries and mushrooms continues to be a very popular activity in city forests. Although mushrooms and berries are primarily picked for own use, some commercial picking also takes place. Sometimes mushroom picking is encouraged by restaurants. Copenhagen restaurants, for example, regularly place advertisements in newspapers that ask people to bring them Porcini mushrooms (*Boletus edulis* Bull.:Fr.). In England, a conflict arose when 'mushroom poachers' started operating in the woods of the Wimbledon Common. Although those caught denied it, wardens suspected that the poachers were selling much asked after fungi to London city restaurants. The Board of Conservators of Wimbledon and Putney Commons opted for a zero tolerance policy against mushroom picking, with intensive patrolling by rangers and hefty fines (Lusher, 2005).

Mushroom and berry picking is seldom a matter of survival, unlike in the days when city forests provided subsistence. In many developing countries, however, people are still dependent on their local forest. One example is that of Ayer Hitam Forest Reserve in industrialising Malaysia. This forest near Kuala Lumpur is under high pressure from development. It is still used by an indigenous community that extracts edible and medicinal plants from the forest (Konijnendijk et al., 2007). The latter are also receiving renewed attention in Europe, where the health-promoting effects of various forest products are being studied, as described in Chapter 9 'The Healthy Forest'.

Although the protection of drinking water resources is generally considered a forest service rather than a product, this function needs to be stressed here. Many countries and cities in Europe, North America and elsewhere manage forests for protecting their drinking water resources. The Swiss city of Basel, for example, obtains about half of its drinking water from the Langen Erlen reserve, a peri-urban forest area of only 94 ha (Bader, 2007). The Vrelo Bosne area near Sarajevo is still the main source of drinking water for the city (Avdibegović, 2003). In some cases, clean water is a main objective of afforestation programmes. This is true for Denmark, for example, where studies have shown that water levels below forests may be lower, but the water is of a higher quality, particularly in the case of broad-leaved forests (e.g. Rasmussen & Hansen, 2003).

Timber, energy wood, non-wood forest products such as mushrooms and game, as well as clean drinking water – the Fruitful Forest is still very much alive today, albeit as part of multifunctional city forests that also provide a wide range of services.

Chapter 5
The Forest of Power

Power has been defined as the capacity to control or change the behaviour of others (Ellefson, 1992). In a classic sociological sense, power relates to the probability that a person or group can assert his, her or its own will in a social relationship, despite resistance (based on Krott, 2005). These definitions seem rather straightforward, but power is a very complex phenomenon (e.g. Mitchell, 2002). Foucault demonstrated, for example, that power is omni-present and not just enforced top-down (e.g. Crampton & Elden, 2006). It is about taking action upon the actions of others, in order to interfere with them. Power is often considered as being 'negative' (e.g. when dictators enforce their will upon people), but it does not have to be repressive and in fact it can also be highly productive. Without power, for example, governments would have difficulties to maintain law and order. Foucault introduced the concept of 'biopower', which among other looks at space as a vital part of the battle for control and surveillance of individuals (Crampton & Elden, 2006).

According to Muir (2005, p. 237), "[i]mages of landscapes have been manipulated and employed in polemical manners for centuries." Landscapes exert subtle power over people, eliciting a broad range of emotions and meanings that may be difficult to specify (Mitchell, 2002). Jacobs (2004) speaks of three different perspectives of landscape, each connected to a different mode of reality, namely physical, social and inner reality. The social reality of landscape includes the aspect of power. Landscapes can be considered as 'powerscapes' which comprise a system of norms that regulates how members of a particular society are required to behave with respect to the landscape. The expression of power in the landscape is a manifestation of law, prohibition, regulation and control. The landscape is a medium of exchange of power, it signifies or symbolises power relations, and it is also an instrument of cultural power (Macnaghten & Urry, 1998; Mitchell, 2002).

The city forest is a powerscape. As cultural landscape and part of the urban environment, it reflects changes in power structures and relations, as for example mentioned by Jones and Cloke (2002), Garside (2006) and Lawrence (2006). The greening of Europe's cities has been "one of the most important, widespread and also controversial of modern urban developments", as reflected in public discourses (Clark & Jauhiainen, 2006, p. 1). The variety and fluidity of green space developments reflect the vital importance of the political process. The history of green space – including city forests – is a mirror of social relations (also Rotenberg, 1995).

C.C. Konijnendijk, *The Forest and the City: The Cultural Landscape of Urban Woodland*

Power has had a major impact on the very establishment of city forests, their ownership, their survival over time, their design and – not in the least – their use. The situation of city forests near population centres may seem to suggest high levels of accessibility and public use, yet as this chapter describes, power has had a distorting influence in this respect. The many benefits of city forests described in this book have not been evenly distributed between different stakeholders, as those in power have been able to claim most uses and benefits. City forestry history is also a history of changing power relations.

Presently, urban green space such as city forests is considered a public space and a place for all (Dings & Munk, 2006). This chapter will demonstrate that it has taken quite some time before this situation arose. The following section shows how the societal elite have, over time, enforced their power through and over city forests. This excluded the masses, but also helped conserve city forests in times of growing pressures. Next, power aspects are explored by looking at the military uses of city forests and the consequences of wars. The links between democratisation processes and city forests are subsequently described, followed by a discussion of how rulers and authorities have turned to city forests to raise their prestige (or that of their city or state). In the final section, links between city forests and environmental justice are explored.

From Wooded Commons to Elitist Prestige

This section introduces how over time, small groups of influential (elite) actors managed to appropriate the ownership and use of city forests. When looking at urban green space over time, the class dimension has always been lurking at the background (Clark & Jauhiainen, 2006). As described in the previous chapter, initially many wooded areas near cities were so-called commons, where peasants had traditional user rights (e.g. Westoby, 1989). However, as discussed, during most of European history, power over city forests and their use was in the hands of a small elite, from aristocracy and clergy to rich merchants. City forests gradually became statements of power (e.g. Lawrence, 1993, 2006).

City forests really started playing an important role in cultural and political life when they became hunting grounds for the King and the aristocracy. Hunting domains not only provided these people with recreation opportunities, they also raised their prestige and helped reinforce power relations. Schama (1995, pp. 144–145) writes about the political importance of the hunt in medieval England: “At this distance it is hard to imagine how vast areas of the country could have been annexed simply to protect royal recreation [...]. But for a warrior state, the royal hunt was always more than a pastime, however compulsively pursued. Outside of war itself, it was the most important blood ritual through which the hierarchy of status and honor around the king was ordered. It may not be too much to characterize it as an alternative court where, free of the clerical domination of regular administration, clans of nobles could compete for proximity to the king.” In his historical account of the Bois de Vincennes of Paris, Derex (1997b, p. 21) confirms that the hunt is a "façon d'affirmer la superiorité du roi" (a way of asserting the superiority of the king). By having sufficient meat and food on the table, and having as many people as possible eat well, the king could confirm his power.

The prevalence of hunting use of forests near cities influenced their use (often commoners were gradually shut out) and the way they looked, as also described elsewhere in this book. During the Baroque period, for example, star forests became popular elements of hunting domains, as they optimised conditions for hunting on horseback, allowing for easy orientation. A large number of today's city forests have had a history as hunting domains, from Bois de Boulogne to Tiergarten, from Castelfusano to Djurgården. Different types of hunting were carried out by the aristocracy, from parforce hunting to falcon hunts. The clergy also enjoyed hunting privileges, as the example of Highgate Wood near London indicates. During the Middle Ages, this wood was the private hunting estate of the Bishop of London (City of London Corporation, 2007).

According to Cornelius (1995), hunting was central to the baroque lifestyle of the aristocracy. In 18th century Berlin, Frederick I, King of Prussia arranged many hunts from his hunting castles, as part of a luxurious lifestyle. The 'Kurfürsten' introduced exotic game such as fallow deer and pheasants to their hunting domains, probably enhancing their prestige amongst those who were invited for the hunt, most notably foreign rulers. In 1849, Berlin's Grunewald, allocated as 'Hofjagdrevier' (court hunting domain), was completely fenced off. It held over 1,000 fallow deer by then, i.e. more than 24 animals per hectare (Cornelius, 1995). The Forstenrieder Park near Munich has been a hunting domain for centuries. During the 18th century, the ruling aristocracy established four hunting lodges in the area. When Bavaria became a kingdom in 1806, the Forstenrieder Park became the hunting domain of the Kings. Large and prestigious court hunts were organised, for example for the benefit of visiting dignitaries like Napoleon I. Conflicts between the King's game keepers and poachers were often severe, leading to casualties on both sides (Ammer et al., 1999).

As said, the aristocracy took to closing off their hunting domains for private pleasure when populations grew and demands for timber and agricultural land increased. In England, 'forests' (as judicial term, allowing the King to fence off areas) were imposed on large areas of the country as early as during the Middle Ages (Schama, 1995; Nail, 2008). These were policed at the king's pleasure, which led to conflicts with the original communities of forest users. Moreover, in times of war and crisis, the King could turn to the forest to raise funds. Richard I of England did so, for example, in search of funds for the forthcoming crusade. He removed from the Forest Law large areas in return for cash, thereby opening up these areas for settlement and agriculture (Nail, 2008).

Wooded and other commons were brought under forest law as well, or sometimes privatised. In France, the 'Code Colbert' (of 1669) led to a power struggle between the rulers and peasants, as this new policy stated that a quarter of all communal woodlands were to be set aside for protected timber (Schama, 1995). In Germany, the so-called 'Bannwald' was reserved for the (hunting) pleasures of the aristocracy and the church (Lieckfeld, 2006).

Although the closing off of these hunting domains excluded large groups of people from using them, there was also a positive side to this expression of power. The strong protective measures taken helped conserve several of these hunting domains until today. At a later point in time, as discussed in Chapter 6 'The Great Escape', many of these hunting domains were opened to the public for recreational purposes, although still not everyone was welcome from the beginning. The better-to-do did not want to be exposed to poverty and misery whilst embarking on leisure activities (Hennebo, 1979).

After royalty and aristocracy, and the clergy who often also owned forested land near cities, a new group of actors emerged with the rise of a rich merchant class in Europe and elsewhere. Especially from the Renaissance onwards, the emergence of these influential merchants changed traditional power relationships. They started to pursue some of the luxury of the aristocracy, such as pleasant environments to live and spend their leisure time. Often heavily-wooded estates were created, inside and more often near cities and towns (e.g. Kuyk, 1914; Lawrence, 1993). The Renaissance villa with its surrounding park, with prime examples such as Villa Borghese in Rome and Boboli Gardens in Florence (Fig. 5.1) provided popular inspiration for these estates. They had a productive function, but recreation and prestige were also very important objectives. Estates were also mostly closed to the general public. In Russia, for example, the gardens and estates near cities such as St. Petersburg and Moscow, owned by the upper

Fig. 5.1 Renaissance estates such as Boboli Gardens in Florence did not only provide a pleasant living environment to their owners. They also helped to raise the prestige and influence of their aristocratic and merchant owners (Photo by the author)

class, were especially used for recreation and prestige purposes. Estates were seen as a 'micro-cosmos' of their own, catering for all of the residents' needs (Jansen, 1995).

Although a larger group in society thus extended its power over forests and green areas in and near cities, greening remained in the hands of a relatively small elite. This was still the case when the forest became part of the Romantic longing to get 'back to nature'. 'Le spectacle de la forêt' (the spectacle of the forest) remained reserved to the bourgeoisie (Kalaora, 1981). As shown in Chapter 8 'The Wild Side of Town', the elite also 'claimed' the wilderness and the first national parks for their leisure experience (Cronon, 1996b; Spirn, 1996); only very few working class people had the opportunity to travel to places like Yosemite National Park in California. However, city forests slowly did become part of democratisation movements and the spaces formerly closed-off to the public gradually opened. President (and later Emperor) Louis-Napoléon Bonaparte of France for example, rode the wave of a liberalisation that swept throughout Europe during the mid-19th century. In 1852, as part of his works to 'green' Paris and create more public green spaces, Louis-Napoléon had the Bois de Boulogne transformed from a royal hunting park into a landscape-style park with meadows and curving paths through the woods (Derex, 1997a; Lawrence, 2006). Works like there were carried out by Adolphe Alphand and overseen by 'prefect' Baron Georges-Eugène Haussmann. The Bois the Boulogne became linked to the city by the grand avenue de l'Impératrice. Clearly, enhancing prestige and reflecting power was also on Louis-Napoléon's mind. As a novelty, parts of the cost of the project were recuperated from sale of property surrounding the renovated forest park. Similar transformations took place elsewhere in Paris (e.g. in the Bois de Vincennes) and throughout Europe.

During the late 19th century, even though many city forests and parks were open to the public, their use still very much indicated class differences and prevailing power relations. Visitors to the various 'hunting forests' near Paris, for example, were divided in space and time according to their socio-economic status (e.g. Derex, 1997a, b). While the Bois de Vincennes had rapidly developed into a popular recreation area for all, Fontainebleau long remained the primary place of recreation for the 'social and cultivated' elite (INRA, 1979). However, the development of the railroads did lead to a certain 'democratisation' of places like Fontainebleau and soon the masses of Paris came to the forest during weekends and holidays. The Rambouillet forest, however, remained exclusive, being reserved to the hunts of emperors and presidents (Widmer, 1994).

Today, 'elite city forests' continue to exist. Some woods in and near cities are privately owned, although many of them have been (at least partly) opened to the public, for example because of public subvention schemes offer much-needed financial support to forest management. City woods and parks are typical venues of the residences of those in power. The Italian presidents reside in the Castelporziano forest park near Rome, members of the Belgian royal family live in the Zoniënwoud and the Irish presidential residency is located in Dublin's Phoenix Park (e.g. Konijnendijk, 1999; Nolan, 2006). La Lanterne, a hunting lodge on the domain of Versailles, is the official country residence of the French prime minister, while the castle of Rambouillet has had the status of presidential palace since 1870 (Connaissez-vouz les résidences,

2005). In the forests around Russian cities like Moscow and St. Petersburg, the rulers and 'new rich' have been building their new 'palaces'.

Hunting has remained associated with elite use of city and other forests, as described by German journalist Lieckfeld (2006) in his much-debated book 'Tatort Wald'. The author argues that upper-class hunters and leading German foresters formed a pact to hold on to elitist hunting traditions, as incorporated in 'trophy hunting', a form of hunting where the deer with the largest antlers is to be shot by the most important person. This so-called 'waidgerechten Jagd' was promoted by people like field marshal Hermann Goering during the era of national-socialism. As a result, game pressures in most German forests were much too high, even far into the 20th century. Generally accepted policies towards closer-to-nature forest management could hardly be realised because of this, until finally – during the late 20th century – sufficient opposition was raised amongst policy-makers, experts, nature conservationists and foresters. That the prevalence of (elite) hunting was not only restricted to Germany can be derived from articles such as that of Hellinga (1959), who complains about high game pressures as a major threat to Dutch forestry. Authorities of cities like Apeldoorn complained in the media about neighbouring forest areas being closed off as 'deer reserves', thus limiting public recreation opportunities.

Military Forests

When talking about city forests as powerscapes, the role of the military and of military considerations in city forest history is so important that it deserves a section of its own. During the Middle Ages, for example, timber from city forests was important both for ships and defence works and to finance war efforts. Efforts of the English Kings to take out territory from under the Forest Law and sell it to generate funding for financing war efforts have been mentioned before. A related case is that of the Haagse Bos, a woodland in the Dutch city of The Hague. Ruler William of Orange wanted to cut the forest during 16th century to finance his war efforts against the Spanish. But the magistrate and citizens of The Hague protested against this, as they wanted to keep the forest for their own use. The forest had already suffered when the Spanish had cut one-sixth of all oaks in the Haagse Bos to support their attack on the city of Leiden. The people of The Hague offered a large amount of cash and even melted the church bells to generate more funding. Thus they succeeded in preventing William from cutting the forest. In the Act of Redemption of 1576, probably one of the first real 'city forest conventions', the agreement was put on paper; the Act stated that the Haagsche Bosch would be reserved for the city and its citizens for all times. This Act has been referred to in the political debate as recent as during the middle of the 20th century, when plans emerged to build in the forest (VerHuëll, 1878; Anema, 1999).

During times of siege, city forests often became victims. The Haarlemmerhout, another Dutch city forest (near the town of Haarlem), was a frequent victim of war activities, much to the grief of the citizens. The forest was cleared by hostile troops, for example the Spanish army, for construction wood, fuelwood and obtaining a free shooting range. Time after time, however, the forest was replanted (Brink, 1984). On the other hand, the defenders sometimes decided to cut the nearby forests by themselves. In this way enemies could not approach unseen and a better shooting range was created. Such a 'clear zone' is depicted on many city views painted or drawn during those times.

Thus during wars, but also during revolutions and other periods of social unrest typically fought out in cities and towns, city forests were often destroyed or at least damaged. This was the case in France, where the French Revolution of 1789 had a direct effect on the Bois de Boulogne, which was plundered and almost completely destroyed by the people of Paris. Later, extensive recreation works were carried out (Bonnaire, 1992; Derex, 1997a). French troops stationed in Hanover in the year 1812 were ordered to cut part of the forest, as compensation for war contributions that had not been paid (Hennebo, 1979).

Many of the impacts of the military and war on city forests can be recognised when looking at the effects of World War II. During this war, city forests were often used by the population to illegally obtain fuelwood, as described by Balk (1979) for the Amsterdamse Bos. In Chapter 4 'The Fruitful Forest', further examples of this are given. Many city forests were used for military bases and for defence works (e.g. Balk, 1979; Slabbers et al., 1993; Van Kerckhove and Zwaenepoel, 1994; Cornelius, 1995). The German occupation forces turned to the Haagse Bos for launching their V-1 and V-2 rockets and a tank ditch was dug in the forest as part of the German 'Atlantikwal' (Anema, 1999). The same Atlantikwal led to the over-cutting of the Mastbos wood near Breda, The Netherlands (Caspers, 1999). During the reign of the Nazi's, overexploitation of Berlin's forests was conflicting with long-term forest plans. Protests of foresters could only partly stop the destruction (Cornelius, 1995). Similar cases could be noted all over Europe. After the war, many (German) forests were overcut as part of compensation payments, whilst forests elsewhere also suffered from a poor and undernourished population. About one-third of the Eilenriede forest of Hanover was destroyed during and after the war, for example due to thousands of bombs (Fachbereich Umwelt und Stadtgrün, 2004).

The relations between city forests and the military also have a more positive dimension. Changing military technology had a positive impact on the development of urban green space, albeit not so much related to city forests. When city walls became obsolete as defence works during the early 19th century in many countries, they were often transformed to green spaces and public walks (e.g. Van Rooijen, 1990). Moreover, city forests have often been used as training grounds, as they were ideally situated close to centres of powers, but still offered space and suitable terrain for exercises. A traditional use of Norra Djurgården in Stockholm, for example, was as military training grounds, but this use stopped when new types of weapons were developed (Nolin, 2006). In the 19th century, the Duke of Wellington used London's Epping Forest to train his troops for the fight against the French (Slabbers et al., 1993). In fact, as in the case of royal and elite appropriation, tightly controlled military ownership of city forests has probably contributed to their preservation until today. Military domains were needed near cities and their transformation to other uses was

Fig. 5.2 Military ownership, as in the case of this woodland in Puglia, Italy, has restricted the use of city forests. It has also helped to protect forests, for example from conversion to other land use (Photo by the author)

prevented (Fig. 5.2). Today, most ministries of defence still own large areas of forests and nature areas, also close to cities. Environmental and recreational considerations are increasingly taken into account by the military.

Another more positive military-forest relation is that city forests have played a role as venues for historical events and as symbols of war and peace. As early as the 17th century, The 'Valcberg' Park in the Dutch city of Breda allegedly symbolised peace in times of war, and peace negotiations took place in the park (Dragt, 1996). The forest of Compiègne in France was the site of signing of the armistice between Germany and France and her allies on November 11, 1918. Two trains drew up from each their side. A stone slab in the middle of the forest commemorates this place (Porteous, 2002). 'Peace forests' have been established all over the world, for example in Hiroshima, Japan, where the first atomic bomb was dropped. The Jerusalem Peace Forest, Israel was created during the 1970s to celebrate the taking of the city by Israeli troops during the Six Days War (Hammer, 2002).

Sometimes even wars have had a positive effect on city forests. Holscher (1973) describes the case of Frankfurt, Germany, were the city dump was turned into a recreation centre, using soil excavated during the construction of the city's subway system. This idea emerged during World War II, when rubble of bombed-out cities was moved out to city limits and buried. Thus artificial hills were created and subsequently planted and landscaped. Kowarik (2005) mentions how after World War II, small woodlands emerged on the fields of rubble in many destroyed European cities.

Lehmann (1999) describes yet another link between military and forest. During the 1930s and 1940s, the army was a cultural symbol for the entire German population. In line with this, the forest was seen as an army, with neat rows of marching soldiers. The forest became associated with the 'Heimat', the home country during the Nazi-era. The German oak was also a much-used Nazi-symbol. This heritage led to a rather uncomfortable relationship between Germans and some of the symbolic values of forests and trees.

City Forests and Democratisation

The first part of this chapter has shown that a small elite decided on city forests and other urban green space for most of their history. But over time, the lower classes have tried to claim their stake, or perhaps rather to reclaim the traditional use rights associated with the wooded commons. Moreover, they have turned to city forests and other green spaces as settings for political and other protests. In this way, city forests have acted as 'marketplaces'. Machiavelli said that the market square evoked fear even with kings, as this was the place where protestors would turn and revolution were born (cited by Gadet, 1992). In many city forests and other green spaces, the organisation of public meetings by religious and political groups as part of power struggles has been one of the most contentious issues (Reeder, 2006b). Socialist organisations in Britain, for example, used green spaces for their manifestations. Epping Forest was often the site of protest manifestations and political meetings. Open spaces such as Kennington Common that had been used for political rallies were turned into more 'civilised' parks, thus neutralising their potential revolutionary powers (Nail, 2008).

In countries such as England, public gardens and green areas were also used by rulers and authorities as a means to distract from political activities, for example by providing a 'meaningful' and moral way for workers to spend their leisure time (Hennebo and Schmidt, s.a.; see also Chapter 11 'The Social Forest'). As mentioned by Garside (2006, p. 81) in the case of London: "Planners presented open spaces as agents of personal and civic improvement, bringing uplifting contact with nature, benefits to health and improved social cohesion. They could also enhance London's reputation as national and international capital."

Forests and trees have also been used as, or perceived as, instruments of political oppression. Tree planting has been used in several countries – including Israel/Palestine – to stake claims over contested land. Thus trees themselves can become symbolic of the nature-value of place and thereby turn into the focus of individual and social resistance based around a particular place (Jones & Cloke, 2002). Trees sometimes had important symbolic value in democratisation processes and revolutions, as in the case of the Lombardy poplar (*Populus nigra* L.), which became a much revered symbol of the French Revolution. Its name in French ('peuplier') is similar to the word for 'people' (meaning 'populace'), thus referring to a transfer of power from the aristocracy to the public (Lawrence, 2006).

As said, however, the lower classes have also stood up against the appropriation of city forests by the elite. An early example of the power struggles between citizens and rulers, the 16th century conflict over the Haagse Bos, has already been mentioned. A more recent example is that of 19th century Berlin, where the city's Magistrate and people confronted the Prussian State over the use and ownership of the forests surrounding the city (including the Grunewald city forest). The Prussian State was accused of letting production purposes determine management and of even planning to sell forests to developers, while city authorities and the people of Berlin wanted to reserve the forests for recreation. Two local newspaper, the Berliner Tageblatt and Berliner Volkszeitung, managed to collect 30,000 signatures for a protest against the partial selling of the Grunewald to developers (Cornelius, 1995).

Another famous example of 'class fights' over city forests and other open space is the battle for the commons in Britain, as described in detail by Reeder (2006b, also Amati & Yokohari, 2006, 2007). Privatisation of the commons had been going on for centuries, for example by the aristocracy and later also bourgeoisie, to ensure their own hunting and other interests. Later, the private (often economic) interest of wealthy residents and estate owners led to the (partial) fencing off of traditional common lands such as Epping Forest and Hampstead Heath near London. Various conflicts over commons arose during the second part of the 19th century due to 'manorial aggrandisement' (i.e. the extension of private estates). In 1854, the population of London protested against the fencing off of properties in Epping Forest. So far, law had been based on the assumption that open, undeveloped land was a waste and its economic development in the public interest. Commons were often being regarded with suspicion by those in power as a refuge of gypsies and a gathering place for dissidents.

The setting up of a Commons Preservation Society in 1865 was the turning point in attempts to change the perceptions of society's relationships with commons. Reeder (2006b) sees this as giving rise to the first 'green' organisation in Britain, in an interplay of local and metropolitan interests. Actions to preserve the commons and keep them out of the hands of economic and private interests were sometimes radical, such as massive trespassing by the so-called Commons Protection League, whilst more often battles were fought out at courts of law. In the case of London, the preservation argument challenged old assumptions and gave priority to the maintenance of commons as open spaces in the interests of Londoners as a whole. Powerful economic organisations such as the City of London Corporation (CLC) threw their weight behind preservation of the commons and became responsible for areas such as Epping Forest (e.g. Green, 1996) and Hampstead Heath (A history of, 1989). One may wonder why an organisation such as the CLC, which mostly had an economic mandate, was concerned with the commons. According to Reeder (2006b), this had much to do with fiscal arguments, as well as with raising its own political standing. The Metropolitan Commons Acts of 1866 and 1870 had handed more control by London's metropolitan government.

London's Parks and Open Spaces Committee wanted to make the commons accessible to a range of users. This should be seen in the light of English public park history, which had great emphasis on regulatory aspects. Parks were viewed as moral preserves whose controlled environment was symptomatic of the greater regulation and surveillance of city spaces during the Victorian period (Reeder, 2006b; Nail, 2008). The commons would have to be tamed, without making them too park-like. Customary uses such as rights of pasture would have to be balanced with urban leisure uses. Anti-social behaviour and public meetings would not be permitted.

The commons did play an important role in the 'democratisation' of the use of open space near cities. "London's parks and commons were an 'other space', a refuge from the crowded streets of the city, promoted as 'green islands', even 'fairylands' [...]. But they were also part of the multiple encounters and rhythms of city life as people from different age groups took them over for different kinds of activities." (Reeder, 2006a, p. 51).

Increased public access to London's wooded commons also had its downsides, as captured in the words of Muir (2005, p. 61) in the case of London: "When elitism and aristocracy had dwindled, the preservation of woodland on the doorstep of Europe's most populous area was a nightmarish task". In their account of the establishment of the London Greenbelt, Amati and Yokohari (2007) also highlight the important role played by private land owners in conserving open spaces near London.

The Commons Preservation Society and Commons Protection League are examples of new actors that emerged in the urban green space discourse. These were associations that mobilised and articulated the views of ordinary city dwellers. Still, there were not 'proxies' as such of the opinion of ordinary city dwellers due to the high degree of networking, professional vested interests and links with local authorities they involved (Clark & Jauhiainen, 2006).

Other groups of actors that gradually extended their power over city forests were city authorities and industrialists. Starting from the 19th century, availability of urban green space to all, and increased focus on the needs of the working classes, became a major concern of these actors, who benefited from healthy workers and a satisfied population. In Helsinki, for example, the availability of green spaces also in working class neighbourhoods became a worry of authorities (Clark & Hietala, 2006). In Sweden, companies such as Ericsson and Marabou established parks around their factories, to be used by their own workers as well as by the general public (Nilsson, 2006a). In the process of providing recreational green space to the masses, remaining forests were often transformed to public parks.

Many cities adopted norms for minimum amounts of public green space. In the Soviet Union and other communist countries, urban planning stressed the importance of public green space available to all. Extensive green structures were developed, new building projects were supplied with ample green and large parks were created, such as the so-called Victory Parks after World War II, known as the Great Patriotic War in Russia. Cities such as St. Petersburg and Moscow were allocated a large surrounding 'forest park zone'. Unfortunately, the massive greening schemes were often carried out rather sloppily, without much thought being given to sustainability and future maintenance. Planting was frequently done in extremely careless way, "so that the residents of the newly built housing estates had to take care of their parks themselves during 'subbotniks' and 'voskresniks' (voluntary get-togethers of city residents for area improvement or cleaning – usually announced on Saturdays or Sundays, hence their names.)" (Anan'ich & Kobak, 2006, p. 267).

Interest and community groups also continued to extend their influence in decision-making processes affecting city forests. Wolschke-Bulmahn and Küster (2006) describe progressing public involvement in the management of the Eilenriede forest of Hanover, resulting in the establishment of a special advisory council (Eilenriedebeirat) in the 1950s. During the past 50 years, this user council has advised city authorities and forest managers over various issues. Moreover, it has raised its voice when new roads, noise, new housing and the like came to threat the forest. Many other city forests today have their own 'user councils' or 'friends' groups. The Vrienden van het Mastbos (Friends of the Mastbos) of Breda, The Netherlands, successfully protested against certain silvicultural activities by the state forest service (the organisation managing the Mastbos). Collaboration then became more constructive and the Friends were engaged in several aspects of forest management (Caspers, 1999). Also in The Netherlands, the Vrienden van Sonsbeek (Friends of Sonsbeek) in Arnhem help the city with information and environmental awareness activities, for example through its information magazine (L. Broer, 2007, personal communication).

City Forests and City Image

At the beginning of this chapter's, we have seen that prestige has been an important factor in the development of city forests and other green space. Prestige of kings, the aristocracy and other people in power was realised through hunting areas, beautification and grand greening projects and schemes – see for example Rotenberg (1995) in the case of Vienna. Rulers and later also city governments have had an interest in creating attractive, prestigious urban environments, to raise their own status, enhance their power and also to attract rich tax payers and businesses. This is definitely what mayor Jules Anspach of Brussels had in mind when he let the elegant, tree-lined Avenue Louise connect the city of Brussels to the Bois de la Cambre (Terkamerenbos) in the 1860s. Moreover, the Bois de la Cambre itself gave the capital of the young country of Belgium a prestigious green area where riders and carriages could parade (van der Ben, 2000). This is very similar to what present-day authorities such as the former Greater Copenhagen Authority have in mind when they state the aim to create an urban region with 'Europe's best recreational landscape' (HUR, 2005). In a time of increased globalisation, competition for money and political attention is not only national, but increasingly also international (Clark & Jauhiainen, 2006). An extensive green structure can help cities move up the ladder of political and economic attention (e.g. Levent & Nijkamp, 2004). City forests can be an important part of an attractive and prestigious green structure.

In his article about the role of public space in the post-industrial city, Burgers (2000) provides an interesting context for the 'prestige' role of city forests and other green space. According to him, there are several key 'types' of public space. 'Erected space' concerns landscapes of economic and administrative potency. Status and prestige are in focus in architecture and design, developed as shared interest of the municipal administration and the commercial sector. Local activities are to be staged internationally. 'Displayed space' relates to landscapes of enticement and temptation. It is all about city marketing and city promotion, stimulating the recreational aspect of a city in the widest sense of the word. Public spaces act as meeting places, especially for urban singles. Cities bear growing resemblance to amusement parks – think of Kotkin's (2005) 'ephemeral city' mentioned in Chapter 1. In 'Exalted space', an erotic landscape of entertainment and ecstasy comes to the fore. Different parks and other spaces offer a variety of sensual pleasures. Exalted space also allows for venting of hostility and for rituals of voicing solidarity (think: the local football stadium). 'Exhibited space', finally, offers landscapes of reflection and elevation. The city becomes a kind of museum, as spatial borders of museums are fading. Focus is on the presence of cultural goods, complexes and settlements, which are all objects of city holidays. Sections of urban space become 'mummified' and listed as protected.

City authorities and private speculators started to recognise the economic possibilities of green space such as city forests during the 19th century. New housing areas were to be developed, and housing in or near green areas came at a premium, attracting richer citizens (and thus taxpayers). In Berlin, for example, large parts of Grunewald were used for housing at the turn of the century. Villages on the eastern border with the city were later incorporated into Berlin. Villa parks were established on the nicest locations, near the water (Slabbers et al., 1993). In 19th century St. Petersburg, many rich citizens, artists and writers built a 'dacha' (summer house) in the forest areas north of the city (Jansen, 1995). Parts of Arnhem's woodland parks were also used for new housing (Kuyk, 1914), as were parts of the city forests of Paris, the Bois de Vincennes (e.g. Derex, 1997b) and the Forêt de Saint-Germain. In the case of the latter, its western rim was reserved for the Maissons Lafitte, the plan for which was derived from the baroque structure of the forest (Slabbers et al., 1993). First concerns about preservation of the Hampstead Heath of London during the 19th century were primarily fed by owners of manors and houses in the area, who wanted to preserve the agreeable surroundings and assured property values (A history of, 1989).

But not only the (partial) conversion of forests and green areas for housing was undertaken. City authorities also saw that attractive green areas could help keep or attract citizens. In 19th century Breda, The Netherlands, city authorities spared cost nor trouble to develop the Valkenberg woodland park into a city park that would be a real 'jewel of the city' (Dragt, 1996). Many more city forests underwent a similar transformation, being redesigned and made into centres of public entertainment.

Today, various businesses take into account the local environment when making the decision to settle in a new area. Analysis of new urban park development in selected British towns indicated that local business and development were boosted when the local environment became more green and attractive (CABE Space, 2005). Interviews with managers of companies based at Dutch 'green' business parks showed that localisation choices are often influenced by the presence of an attractive green environment (Böttcher, 2001). Managers see the benefit of a green setting in terms of representative quality and image towards customers, local residents and citizens. Moreover, surrounding green space can have a positive influence on personnel (see also Chapter 9 'The Healthy Forest'). In China, a good investment climate in firms seems to be linked with a good environment for people. Cities with a better investment climate score higher on environmental measures, including green space per capita (Dollar, 2006). Retailing is also affected. Work by Wolf (e.g. 2003) in the United States has indicated that shopping areas become more attractive to people when they are greener.

However, trees and woods are not only good for commercial activities. The role of city forests and parks in creating prestigious living and working environments has been quantified in economic terms, for example by Luttik (2000), CABE Space (2005), NUFU (2005) and Tyrväinen et al. (2005). Nearby green space can have a significant effect on property values. People consider having green space nearby as an important argument for buying a house, together with factors such as nearby schools, shopping opportunities and public transportation, as well as obviously various house characteristics (e.g. Van Herzele & Wiedemann, 2003). Studies in North America and the United Kingdom have shown that house prices are between 5% and 18% higher where property is associated with mature trees (NUFU, 2005). A recent study in The Netherlands found a price premium between 4.5% and 15% for houses with a view of parks, water and open landscape (Gerritsen et al., 2004). Increases of several percents or more have also been noted in Finnish cities by Tyrväinen (1999). Effects can be very dramatic; properties bordering new urban parks in the United Kingdom were found to have price premiums as high as 34% (CABE Space, 2005). However, price premiums are not consistent and depend on local conditions.

A very interesting case is that of Bold Colliery, a former mining community at the edge of St. Helens, Merseyside, England, that became a scarred and degraded area after mining was stopped in the 1980s. However, afforestation and landscaping of the area was started by The Mersey Forest and Groundwork, St. Helens, and more than 500 new houses were built. The area became popular once again, leading to significant increases in house prices. In a report to the Forestry Commission, the so-called District Valuer stated in 2004, that the community woodland in the area 'directly and uniquely' had enhanced existing properly values in the surrounding area by 15 million GBP (District Valuer's Report, 2004).

City forests can also play a role in defining the specific culture and image of a city (Fig. 5.3), just like Central Park helps to identify New York (Treib, 2002). A better image makes a city more competitive, thus expanding its political and economic influence. Djurgården in Stockholm, representing a typical Swedish rural landscape in the heart of a capital city, even helped in developing a nation's identity during the 19th and early 20th century (Nolin, 2006).

Due to the rise of the leisure economy, tourism is of increasing importance to many cities (Jansen-Verbeke, 2002). Forest and green space can contribute to promoting tourism, as main attractions or – more commonly – as attractive 'setting' or backdrop. As early as in 1929, Stockholm guidebooks stressed the environmental attractions of the city, including Norra Djurgården, where nature was said to display

Fig. 5.3 Tiergarten in the heart of Berlin has provided a prestigious setting to the city and its rulers. Just like other Berlin landmarks, such as the Brandenburger Tor, Tiergarten is part of the city's identity and attractivity (Photo by the author)

her beauty (Nilsson, 2006a). Helsinki's city guide from 1936 also stressed the city's 'green image'. Its numerous parks and plantations, its sea breezes made Helsinki into a real health resort (Lento, 2006). The Parque Amazónia in Belém, Brazil is a wooded park of about 7,600 ha close to a city of 1 million people. One of the main objectives of this tropical forest park, promoted as Gateway to the Amazon, is to attract tourism (Parque Amazónia, Belém, 2006). The tourism sector has been able to capitalise on some of the positive effects urban green space. Hotel rooms in Zurich that have a pleasant view, for example on a park, are most frequently occupied. Moreover, they are rented out at a surcharge of on average 20% (Lange & Schaeffer, 2001). In some cases, green-space management organisations can also benefit directly from commercial recreation and tourism in their areas.

Tourism and city marketing at large can benefit from various types of events held in city forests and parks as well. Events in city parks and forests can range from games and sports competitions to exhibitions, festivals, parties and music concerts. German city parks have attracted up to 25,000 participants during parties and concerts (Gehrke, 2001). Some events are thematically linked to the green space, but often there is no link at all.

City marketing through green space can take many forms. Cities have turned to getting their green spaces and nature nationally or internationally recognised, for example. City forests have also been included in forest management certification schemes such as the one provided by the Forest Stewardship Council (Konijnendijk, 1999). National city parks have been established in cities like Stockholm (Djurgården) and Hämeenlinna, Finland (e.g. Schantz, 2006) – see Chapter 13 'A Forest for the Future' for more information about this. 'Ordinary' national parks or nature reserves of national importance are found within or just outside cities such as London, Sarajevo, Vienna, Warsaw and Moscow. Woodlands are prominent parts of these parks. The Wienerwald near Vienna was granted the status of Biosphere Reserve by UNESCO (Umweltdachverband für das Biosphärenpark Wienerwald Management, & Österreichische UNESCO Kommission 2006).

In some cases, cities can use their long-standing links with forests as marketing opportunity (e.g. Konijnendijk & Flemish Forest Organisation, 2001). Many cities and towns have names that relate to forests, such as the Dutch cities of 's-Gravenhage (meaning Wood of the Counts, now The Hague) and 's-Hertogenbosch (meaning Forest of the Dukes, now Den Bosch). A study by Fei (2007) in the United States shows an abundance of place names relating to trees and woods. Pinegrove or Pine Grove, for example, appears more than 1,500 times across the USA. In total, more than 36,000 place names with apparent associations with 19 tree species or species groups were identified in the city.

City Forests and Environmental Justice

According to Di Chiro (1996), environmental problems are most often borne by the poor and marginalised. We have seen in this chapter that the elite have been able to extend their power over others, also in city forest issues. They have been able to claim city forests or at least reserve some of their uses in order to deal, for example, with the less pleasant aspects of living in urban areas. City forests provided them with space, peace and quietness, clean air and the opportunity to be away from masses of working class people that did not behave like them. Di Chiro also mentions how nature conservation organisations have been criticised being biased towards only part (elite, middle class) of society. This had resulted in resentment amongst parts of society. Elite notions of nature have become a tool of oppression. This was the case for villagers and peasants being pushed out of forests that became hunting estates. This is still the case today, for example amongst ethnic minorities in North-American cities.

In the specific case of urban forests (i.e. all tree dominated urban vegetation, including city forests), Heynen et al. (2006) have noted that in the United States and Western Europe, the urban poor and minorities are underserved by the urban forest, meaning that they have less access to green spaces where they live and work (cf. Barbosa et al., 2007 in the case of Sheffield, UK). This is acknowledged by Conway and Urbani (2007), who state that lower levels of vegetation are consistently found in poorer neighbourhoods. In the case of South Africa, McConnachie et al. (2008) found more and better quality green space and a higher diversity of woody plants in wealthier areas of towns. Poorer residents often lack the resources to enhance the urban forest on their own property, stressing the need for public policy and action to ensure equal access to the urban forest across a municipality.

Community involvement in city forestry and other efforts related to the local environment are crucial for providing environmental justice, as other parts of this book also show. Di Chiro (1996) states that social and ecological sustainability should go hand in hand if we want to maintain or create liveable cities.

However, community involvement also has a power dimension, as new territorial boundaries are drawn up. Well-meant community efforts to protect 'their' city forest or green space, which are often criticising 'those in power' for not sufficiently taking into account the community's wishes, can backfire. Jones and Cloke (2002) describe the case of the wooded Arnos Vale Cemetery of Bristol. A local community group was set up to prevent further abuse of the run-down cemetery, where crime and vandalism had become rampant, partly because of lack of supervision and social control. The group installed a regime of monitoring and surveillance, which then resulted in other visitors feeling unwelcome to use the area. Overtones of power and morality had come into the picture.

In the case of young people, Thompson et al. (2006) mention that public attitudes that cast them as problem of threat can lead to marginalisation and social exclusion, also where woodland use is concerned. When asked for barriers they perceived in use of 'wild adventure space', the youths themselves mentioned fear of other teenagers and gangs among the obstructing factors.

These examples show that power relations are still important in modern city forestry. They influence, for example, social accessibility of city forests. Not all inhabitants have equal opportunities for using city forests, for example, for recreational purposes. This is problematic, as it has an effect on people's quality of life and health (see Chapter 9 'The Healthy Forest'). On the other hand, local residents have been trying to claim their rights, getting involved in various efforts to become more influential in decision-making with regard to city forests. More participatory planning and management have emerged from this, as discussed in Chapter 11 'The Social Forest'. That these efforts sometimes lead to conflicts is described in the chapter 'The Forest of Conflict'.

Chapter 6
The Great Escape

City forests have always taken a rather special place within urban societies. On the one hand they are an integral part of them, on the other hand they offer – at least the illusion of – a possibility for escape and retreat. In line with Tuan's (2007) distinction introduced in Chapter 1, city forests are both safe, familiar 'places' and unpredictable, exciting 'spaces'.

In this chapter, city forests are analysed as 'antidote' to the routines of urban life, as a contrast to more urban settings. The chapter looks at how city forests have become popular as (recreational) escapes, enabling a wide range of activities that became popular as societies developed. City forests have helped to offer a 'window to the world', introducing people to the plants, animals, people and cultures of other parts of the world or bygone days. Forest recreation has changed over time, from walking and more sedate forms of nature experience (which are still very popular) to a wider range of leisure activities, including more adventurous ones. Our cities and towns are made up of many different people and obviously these people do not all have the same preferences regarding city forest use. They all have their own way of searching for The Great Escape. The final section of this chapter looks at city forests as ultimate escape, that is, as areas that have been – at least partly – beyond the law and control of the city.

Antidote to the City

The city forest is an interesting and maybe seemingly contradictory phenomenon. City and forest have often been described as opposing ends of a continuum. Where civilisations emerged forests where cleared, and the other way around (Tuan, 2007). In his book 'Forests: Shadows of Civilization', Harrison (1992) discusses the complex relationship between forests and civilisation. Initially, there was only the wild forest, also described as 'the desert' and as brutal nature (INRA, 1979). Harrison compares the rise of civilisation with the clearing of the forest canopy. The clearing got bigger when civilisations developed (and the city can be seen as the ultimate 'clearer') and the forest was pushed to the boundaries. Still, forests continued to play an important role, as reminder and very source of civilisation (see, for example,

Chapter 4 ‘The Fruitful Forest’), as a source of knowledge (see Chapter 10 ‘The Forest of Learning’), and as source of inspiration for artists, poets, as well as common people (see Chapter 7 ‘A Work of Art’). The forests thus provided the opportunity for an ‘escape’ from civilisation, for exploring the roots of humanity. In this way, forests have always provided a kind of antidote to the city and its hustle and bustle. On the one hand, they offered something very different, while on the other hand they were still part of the city as expression of civilisation. From this viewpoint, forests at the urban fringe continue to be of importance as ‘frontier’ area, being part of the city as well as placed outside at the same time (Thomas, 1990).

Lassøe and Iversen (2003) mention that nature has two important roles in modern society. First of all, it is (or should be) an obvious part of everyday life. Secondly, however, nature is a contributor to extraordinary experiences. This is in line with literature that shows that people prefer managed landscapes close to their home, but they also need ‘wilder’ green areas, including woodlands, close to their neighbourhood (Jorgensen et al., 2006). In today’s urbanised society, nature has become an arena for human play and leisure and for escaping from the routines of daily life (White, 1996).

The complex role of woodlands near cities and towns is also discussed by Jorgensen and Anthopoulou (2007). For people, woodlands may constitute a bounded other world, the world of childhood dens, a woodland garden. They may be associated with spirituality and transcendence (see Chapter 2 ‘The Spiritual Forest’). Woodlands often contain highly enclosed environments, paradoxically inspiring feelings of both containment and freedom. They evoke feelings of ‘intimate immensity’ (Bachelard, 1958, cited by Jorgensen & Anthopoulou, 2007). In interviews with 20 Danish forest users, Oustrup (2007) found that the forest is indeed regarded as an antidote to the city, as a contrast to the civilised and cultivated. Still, a certain amount of ‘civilisation’ is also asked for in the woods, as other visitors are expected to behave respectfully. The forest is seen as a place to disconnect and de-stress. The ‘non-controlled’ experience of the forest is crucial to achieve this ‘escape’.

City forests are appreciated as offering – at least to some extent – this ‘otherness’, and as providers of freedom and opportunity to escape from daily city life. On the other hand, they also play a role as a kind of marketplaces (see Chapter 11 ‘The Social Forest’), where urbanites gather, meet and socialise. Although they are not as ‘urban’ as a city square or a city park, city forests are still undeniably a part of the urban environment and of urban life. Some forests, in particular those situated close to where many people live, can even have a role that is very similar to that of city parks. Their use then becomes similar to that of the Amsterdam inner-city parks studied by Gadet (1992), who found that shorter-term, more dynamic uses are becoming more popular. Typical activities are short walks, walking the dog, having lunch, cycling through a park on the way home, and not in the least meeting with friends or observing other people. A park such as the Vondelpark in the heart of Amsterdam is especially popular with the city’s singles, as a meeting point.

Wooded areas can also offer a more permanent escape within urban areas. During the 19th century, for example, the middle classes of New York moved out from the city centre and created attractive green 'villages' in the urban fringe. But these green neighbourhoods were soon incorporated by urban development and the "restless flight to the north", farther away from the city centre, continued (Hiss, 1990, pp. 179–180).

Recreate and Enjoy

Spending leisure time on a busy forest meadow together with other people, as well as hiking alone through a dark and wild forest, are both possible in most city forests. Recreation and enjoyment are central to The Great Escape.

City forests have sometimes been described as the oldest recreation areas, as in the case of the Dutch Haarlemmerhout, a wood in the city of Haarlem (Brink, 1984). During the Middle Ages, public leisure and garden life in Germany and other countries (i.e. for those who could afford to spend time on these activities) in were mostly occurring outside the cities (Hennebo, 1979). Inside the walled cities, public green space was very limited.

During the Middle Ages, but particularly from the Renaissance onwards, city forests gradually started to become 'centres of pleasure'. They offered opportunities for hunting and many other activities. With the rise of a class of merchants that demanded opportunities for sports, games and outdoor socialising during the 17th century, a new – elite – leisure class had emerged (Van Rooijen, 1990; Lawrence, 1993).

Park Valcberg (now Valkenberg, i.e. Falcons' Hill) in the city of Breda, The Netherlands, for example, was developed by Count Henry III (Henrik III) during the 16th century. Henrik was influenced by his travels and impressed by the Renaissance style. An early 17th century engraving of Valcberg shows embellishing and recreational elements such as parterres, a pall mall, a star forest, a rose garden and a fishing pond. Apart from its productive uses, the woodland park was clearly laid out for pleasure, offering a variety of landscapes and experiences (Dragt, 1996).

Initially, the main attractions of inner-city and nearby nature areas were not the forests themselves. The first (16th century) visitors to Fontainebleau did not come to see the dark and dangerous forest, but rather to admire the local castle, or for a festival (Kalaora, 1981). The first 'Waldgaststätte' (forest inn) in Hanover's Eilenriede forest was established in 1681, followed by the further development of a recreational infrastructure (Hennebo, 1979). Even during the 19th century, visitors to London's Hampstead Heath seemed more interested in pleasure gardens, inns and fairs than in the landscape of the Heath itself (A history of, 1989).

Many other city forests and woodland parks – often owned by the aristocracy and formerly used as hunting grounds – became popular places for recreational escape during the late 17th and early 18th century. The Prater in Vienna, for example, a favourite hunting ground of the royal family, was opened to the Viennese in 1766. There were some reservations amongst the upper classes, however, about the behaviour of the 'common' public (Buchinger, 1967). Djurgården, a former royal hunting ground in Stockholm, became accessible to public in the 18th century. Early on it already housed numerous amusements, inns and cafés, as well as large number of private estates (Nolin, 2006).

During the Industrialisation, i.e. for most European countries during the latter part of the 19th century, the modern concept of leisure emerged. The associated new ways of allocating time and the changed perceptions of work gave leisure activities a 'fresh' social meaning (Myerscough, 1974). No longer was there only a (well-to-do) leisure class, but society started to develop into a "leisure society" (Fairbrother, 1972, p. 43). Providing public leisure spaces became a task of private and public actors. Their efforts were facilitated by the presence of existing green spaces, often in (formerly) royal or private ownership (Fig. 6.1).

In France, Fontainebleau became a true 'Parisian promenade' (INRA, 1979) – in spite of its distance of about 60 km from Paris – when the railroad from the capital reached the forest in 1849. Kalaora (1981) speaks of 'trains of pleasure' bringing thousands of Parisians to the forest. The royal hunting grounds in Paris also became popular places to visit. Especially on Sundays and holidays the Bois de Boulogne, for example, attracted many visitors (Derex, 1997a). But still not many people had the same opportunities for leisure, for example because of long working days and limited mobility (Hennebo, 1979). The former hunting domains all had a character of their own, as also mentioned in Chapter 5 'The Forest of Power'. Where the Bois de Vincennes was a place for common people to meet, the Bois de Boulogne was more mundane and attracted the bourgeoisie. Fontainebleau, on its turn, was a place where the social and cultivated elite relaxed and assembled. So the population was divided over the areas according to their socio-economic status. This situation was maintained through social control (INRA, 1979; Derex, 1997a, b).

Hampstead Heath obtained its own railway station in 1860. This, together with the British Bank Holiday Act of 1871 which meant that public holidays now also fell in months favourable to visiting green spaces, led to a dramatic increase of recreational uses by the working classes. One some days, as many as 50,000 people visited the Heath (A history of, 1989). In the period between 1820 and 1878, Epping Forest became of growing importance as public green space for London. Hunting, fairs and sports all remained popular. Some owners of manors located within the area decided to fence off their lands, leading to conflicts with the visiting public (Slabbers et al., 1993).

The forests of Berlin started to become very important for recreation at the end of the 19th century. Recreation was, as in the cases of Paris and London, stimulated by the development of the railroad network. In 1849, Grunewald still was a fully fenced 'Hofjagdrevier' (royal hunting reserve). Its public recreational use would remain limited until the beginning of the 20th century (Cornelius, 1995). Also in Berlin, the recreation facilities in Tiergarten were further extended when a zoological garden was opened in 1844, at the site of the old pheasant garden of Frederick II (Nehring, 1979). In 19th century Gdansk, Poland, excursions in the neighbourhood of the city became popular amongst the 'propertied classes'. People went by coaches available from the city gates, as well as by barge and later railroad (Cieslak & Biernat, 1995).

Fig. 6.1 During the 19th century, recreational visits to city forests became very popular throughout Europe. Attractions included rides in horse-drawn carriages. Visitors to the Vrelo Bosne park near Sarajevo can still enjoy this pleasure (Photo by the author)

Use of the Oslomarka forest by the citizens of Oslo underwent a very similar development. From the second half of the 19th century onward, rather exclusive skiing associations became active. The first café in Oslomarka opened in 1889 (Aalde, 1992). By then, recreation in Scandinavia and elsewhere had become more 'democratised' and leisure time of the different classes increased, as did mobility. Møller (1990) describes in detail, for example, how the various classes of the citizenry of Copenhagen went on a 'skovtur' (forest tour) to Jægersborg Dyrehave.

Places such as wooded Central Park in New York had the clear objective to act as a refuge for the hundreds of thousands of workers, who did not have the resources to spend their summers in the country (Lawrence, 2006). The Park should offer, in an expensive way, similar benefits as a month or two spent in the White Mountains or Adirondacks by the richer part of the population. The role of Central Park as a refuge and a getaway is still clear today from its 'inward' orientation. The Park has its 'back' towards the city, setting it apart and creating the illusion of a larger-scale natural landscape.

Recreational use of city forests increased further during the 20th century. Today, city forests attract thousands of visits per hectare annually – many more than other forests (Konijnendijk, 1999). Most city forests are primarily used for walking, cycling and various sports activities. Experiencing nature and just 'getting away' also score high when people are asked for their motivations to visit (city) forests. The average Dane, for example, is a frequent user of forests, parks and other green areas; especially walking is popular (Holm, 2000; Jensen, 2003; Kaae & Madsen,

2003). Forests attract about half of all recreational visits to nature and green, followed by beach and coast (which account for 28%; Jensen, 2003). More than two-thirds of all Danes see nature as an important part of everyday life and even more see it as contributor to their quality of life (Kaae & Madsen, 2003). Main motives for spending time in nature are to experience nature, find peace and quietness, and strengthen family ties.

The availability of good opportunities for outdoor recreation and the 'green' qualities of the neighbourhood and city have become important factors in people's decisions to settle and live somewhere, as already mentioned in Chapter 5 'The Forest of Power'. Lack of green space and children's playgrounds, for example, was identified as a main reason for people moving out from the city of Leuven, Belgium (Van Herzele & Wiedemann, 2003). In Helsinki, 92% of respondents in city-wide poll found nature to be important (City of Helsinki, 2005). Residents of Barcelona recognise the importance of green spaces for the quality of life in the city. Tranquillity and contact with nature are seen as major advantages of living in the city (Priestley et al., 2004).

Obviously the use of city forests for leisure purposes differs across Europe and across user groups, in spite of the similarities mentioned above. A traditional 'everyman's right' exists in the Nordic countries and parts of Eastern Europe, stimulating regular forest use also off the paths. City forest use may also be locally determined. In the Wienerwald, for example, visits to the forest are often accompanied by a church visit (Bürg et al., 1999). Moscow's National Park Losiny Ostrov is a very popular recreational area, with short-term visits for walking, jogging, picnics and skiing in winter as preferred activities (Sapochkin et al., 2004). Walking is also a preferred activity in the city forests of Berlin. A study by Meierjürgen (1995) indicated that 60% of the people of Berlin go to the forests to walk. However, nature observation (mentioned by 40% of the respondents) and cycling (25%) are also very popular. As also described in Chapter 4 'The Fruitful Forest', hunting remains popular in large parts of Europe. This is even the case for city forests. The forest of Rambouillet, for example, is still the hunting domain of the French leaders (Widmer, 1994).

In spite of the popularity of city forests, studies also indicate that not every city dweller uses the local woods. In some cases people state that they are not interested or have no time. In other instances, people do not visit forests because of fears (see Chapter 3 'The Forest of Fear'), lack of awareness or absence of the means, health or mobility to get to the forest. This means that The Great Escape is not (yet) available to all.

Window to the World

City forests do not only offer people with opportunities to recreate, they also provide a kind of 'window to the wider world' and thus an escape from everyday reality. Arboretums have made people familiar with the trees of far-away continents. Zoological gardens and game parks have done the same with exotic (and less exotic) animals. Exotic game such as fallow deer (Dama dama L.) and muflon

(Ovis musimon L.) were introduced to the hunting domains of Western Europe. Even today, many city forests still host zoological gardens, although in other cases – such as the Haagse Bos – zoos have disappeared (Anema, 1999).

Amusement parks and fairs in city forests have introduced people to artists, arts, foods, customs and so forth from all over the world. The annual 'Boskermis' (forest fair) in the Haagse Bos in the Dutch city of The Hague was very popular with local residents. The event started in 1706, but was stopped in 1729, primarily due to complaints about improper behaviour associated with the fair (Anema, 1999). Open air museums have showed everything from a nation's own cultural history to life in an African village. The 'Exposition coloniale' which took place in the Bois de Boulogne during the early 20th century, for example, included an entire Congolese village. The millions of visitors had a chance to taste food from many different countries (Derex, 1997a). City forests helped to show that there was a fascinating world outside the city, outside the country and on the other side of the ocean. Art festivals, plays and concerts in city forests have also brought the world to local residents.

Today, most Europeans have ample opportunities to travel and experience the world. However, not long ago, travelling was reserved to the elite. American landscape architect Frederick Law Olmsted was one of the people who realised that not everyone had the same possibilities to see, for example, the beautiful landscapes of America's west (Cronon, 1996a). He therefore wanted to bring part of this to regular people's backyard, for example by imitating natural landscapes in New York's Central Park. In a way, similar thoughts are reflected in the plans to develop the Zurich Sihlwald into an urban 'wilderness experience park' (Bachmann, 2006; see Chapter 8 'The Wild Side of Town' for further information). Not everyone has the opportunity to visit 'real' wilderness and therefore wilderness is brought right onto people's doorsteps. 'Wild' city forests are a window to the remaining 'wild world' out there (e.g. BUND Berlin, 2007), or perhaps to an ideal of primeval, untouched nature that no longer exists.

City forests continue to provide a window to the world. In Israel, for example, the population has a wide variety of cultural backgrounds, resulting in many different ways of using the forest. Religious and cultural festivals have been used to make links between cultural diversity and city forests (M. Shaler, 2006, personal communication). Europe's population is becoming increasingly multi-ethnic as well, and city forests can be public places where different cultures can meet (e.g. Worpole & Knox, 2007). Kotkin (2005) mentions how important it is for cities that cosmopolitan attitudes are nurtured and developed. Openness to varied cultures has been one of the driving forces of the most successful cities throughout history.

News Ways of Escaping

The recreational use of city forests and other urban green areas has changed over time and continues to change, for example as a result of the move towards a leisure, information and ecologically-oriented society (Bürg et al., 1999). Pröbstl (2004) mentions that recreation and leisure behaviour in Europe has become more individualistic and less organised. People expect to participate in activities of their choice, at times and locations convenient to their own lifestyle. Leisure activities in Marseilles, France, for example, have changed rather dramatically. Recreational commuting in the form of mass migration to more distant natural areas has boomed (Werquin, 2004). Currently two-thirds of the residents of the city do no longer use the smaller public gardens inside the city for weekend and holiday leisure activities. More active and adventurous forms of recreation are increasingly becoming popular, as illustrated by a study of recreational use of the Wienerwald, Vienna (Bürg et al., 1999). A study in Finland by Pouta et al. (2007), aimed at assessing future developments in outdoor recreation, also points at the decline of more 'traditional' activities (such as berry picking) and an increase in more modern, urban activities, including motorised forms of recreation. More information about new, more adventurous types of recreation is provided later in this section.

National studies of recreational use of forests and nature in Denmark (e.g. Jensen, 2003) have identified several trends that probably also are of relevance to other parts of the world. The number of visits to Danish forest and nature areas increased from the 1970s to the late 1990s, while duration of the average visit decreased. Decreases also occurred in usual group size, transportation distance and duration of travel. The case of the Amsterdamse Bos in The Netherlands can serve as an example of changes at the level of individual city forests. While mass use in the form of sunbathing, swimming and picnicking had been dominating during the 1960s, walking, jogging and walking the dog had become the most popular activities by the late 1980s. By then, almost 75% of all users stayed in the Bos for two hours or less (Bregman, 1991).

A trend towards more nature-based forms of recreation has also been noted (e.g. Roovers, 2005). Nature-based forms of recreation and tourism, defined as primarily concerned with the direct enjoyment of some relatively undisturbed phenomena of nature (Valentine, 1992, cited by Roovers, 2005), have increased rapidly, especially during recent years.

Obviously the popularity of city forests and other green spaces and their use are to a large extent defined by what people think about these green spaces. In Helsinki, most positive opinions related to local green spaces are associated with possibilities for activities and with the perceived beauty of the landscape. Also mentioned in a recent study of a Helsinki city district were 'freedom and space', 'forest feeling', and 'peace and quietness' (Tyrväinen et al., 2007). Visitors of the woodlands of the town of Redditch, United Kingdom most frequently expressed the emotions of feeling happy, relaxed and close to nature (Coles & Bussey, 2000). Hammitt (2002) describes how urban forests and parks are used as 'privacy reserves', as indicated by a study of visitors of city parks of Cleveland, USA. Certain park settings were preferred over others for privacy. 'Reflective thought' was seen as the most important function that privacy served within refuges (Fig. 6.2).

Modern uses of city forests do not only relate to quietly observing nature and finding privacy to contemplate, however. As found by several of the studies listed above, trends are towards more activity and adventure-oriented forms of recreation.

Fig. 6.2 Many city forests offer opportunities for finding peace, quietness and privacy, as here in the case of the Lorettowald in Konstanz, Germany (Photo by the author)

Recent years have seen the development of facilities such as mountain bike trails, activity centres, nature playgrounds, and so forth, often centred in forest areas close to urban areas.

Lehmann (1999) describes how many 'adventurous' uses of forests have emerged in Germany. In the old days, hiking or 'wandering' were considered main activities to 'escape' from daily routines. However, in spite of popularity, many people today think of the traditional 'wandering' or hiking as being old-fashioned. Today, more active forms of recreation, such as mountain biking, are much more popular. Especially with the younger generation, paint balling and (semi-legal) parties are favoured forest uses. Once again, the city forest becomes a place of playing out one's fantasies. City forests like the Bambësch Forest (situated near one of the richest neighbourhoods in the city of Luxembourg) have become activity centres. They offer a wide range of sports facilities and large-scale playgrounds (Millionærer og børns, 2006).

A study by the Dutch Alterra research institute looked at six new forms of forest recreation in The Netherlands: mountain biking, 'struinen' (off-trail walking), 'rubber boot trails', play forests for children and youths (see the next section), 'pole camping' (primitive camping in nature, with a pole identifying the camping plot) and GPS-walks. The use of GPS (Geographical Positioning Systems) is a key element of an increasingly popular activity called 'geo-coaching'. A so-called 'cache' or 'treasure' is hidden somewhere in nature and its location is given on a certain web site. Participants can then use their GPS to search for the cache in a modern type of treasure hunt. The Dutch State Forest Service has started offering possibilities for geo-coaching in its forests (Romeijn, 2005).

Role playing or adventure games, where people dress up as elves, wizards and knights and play out their fantasies, have become rather common uses of many city forests, inspired by, among others, the popularity of the Lord of the Rings movies. A Danish newspaper article describes the popularity of a Copenhagen wood named Brøndbyskoven for this type of activity (Dalum, 2002). A special club, Loriana, organises the role playing games. However, use of the forest for fantasy games was recently prohibited by the local municipal council, as other users had complained about the noise and 'violence' of the games. Thus Loriana's activities were moved to another – in the eyes of Loriana much less attractive – forest. This resulted in the club's 'charm offensive' amongst local residents, but the local council stood by its decision.

Another form of 'wooded escape' is illustrated by the situation in France, where the building of and temporary living in tree houses ('cabanes' in French) has become popular (Darge, 2006). 'Elf dwellings' or wooden fairy-tale houses are built in the tops of trees, offering people a chance to live out their childhood dreams. Commercial enterprises have started building tree houses ranging from small and simple to large and luxurious for various customers. A tree house in the Paris city forest of Rambouillet can be rented to listen to the mating call of red deer. Some people use their tree houses as a 'love nest'.

Heavily wooded theme parks, open air museums and other 'entertainment landscapes' sometimes have the character of city forests, although their use is mostly semi-public, as an entrance fee has to be paid. Lörzing (1992) describes how the creation of theme parks and zoological gardens in The Netherlands has stimulated the development of post-modernistic landscape design. Trees and woods help to create the right setting for entertainment and amusement. The Efteling theme park near Kaatsheuvel, The Netherlands, for example, includes a special 'fairy tale forest'.

Different People, Different Escapes

Many people live in cities and towns and these people have a wide variety of preferences regarding city forests. People's appreciation of different types of recreational landscapes varies, with forests often scoring highest, followed by beaches and coastal areas, and lakes and ponds (e.g. Reneman et al., 1999; Jensen, 2003). Efforts have been made to recognise different groups of forest users, in order to assist forest managers and others to meet varying demands. De Vries and De Bruin (1998), for example, segmented outdoor recreationists in the categories 'the reluctant', 'the family oriented', 'the satisfied', 'the busy' and 'the weary'. This division is based on people's age, lifestyles, social preferences, and the like. In Vienna, researchers identified several 'user profiles' according to typical activities and behaviour. Forest visitor types included, for example, 'the active and mobile'

(23% of all visitors), 'the nature watcher' (21%), 'the social' (16%), and 'the sporty and relaxed' (10%) (Bürg et al., 1999). Picking berries and mushrooms are activities carried out by 43% respectively 30% of the residents of Helsinki (City of Helsinki, 2005). Use of forests differs between groups, but also within a country. Sievänen et al.'s (2003) study of regional differences in recreational use in Finland showed that Southern Finland – which is much more urbanised than the rest of the country – has a higher incidence of use of summer cottages, outdoor swimming, as well as of berry picking.

Among the visitors of Fontainebleau – which attracts an estimated 12 millions visits per years, making it the most visited forest in France – two major groups of visitors can be distinguished: the people of Paris and tourists. The Parisians mostly use the forest for picnics and hiking, while tourists focus more on attractions, such as the castle (Trébucq, 1995). The differences between regular users and tourists are also recognised by Skov-Petersen and Jensen (2007) who distinguish between three major types of forest use (in Denmark), namely daily use, weekend use and tourist use.

One important group of city forest users is that of children and teenagers. Holm (2000) found that 7–16 year-olds in Danish cities use parks more frequently than other age groups. A national survey among Danish 10–14 year-olds identified that 49% of them identified playing as main use of nature (Danmarks Naturfredningsforening & TNS Gallup, 2004). Role playing was mentioned as a main use by 11% of the youths. In contrast, Dutch studies (e.g. Verboom, 2004) have indicated a rather limited use of nature amongst Dutch teenagers. For her doctoral study, Kirsi Mäkinen (in prep) interviewed youths (14–15 year and 15–19 year olds) in Helsinki. Results of the study include that adults generally use green areas more frequently than youths, and that youths also use green spaces differently. Green space use also differed between boys and girls, with boys engaging more in sports activities. Youths and teenagers see city forest as a possibility to escape from parental and social control, as a study in Scotland (Thompson et al., 2006) has shown. The forests offer youths with the opportunity to be amongst themselves and to engage in activities that may be frowned upon by (adult) society.

Child's play has become much debated recently. The mentioned research in Scotland, carried out by the OPENspace Research Centre on behalf of Nature England, for example, stresses the importance of 'wild adventure space' for young people (Thompson et al., 2006). The wild environment can play an important role in meeting the development needs of young people. This was also stressed during a conference on Urban Wildscapes held in Sheffield during September 2007 (Simson, 2007b).

However, wild adventure space is still poorly understood and associated with many barriers, such as safety concerns amongst adults. People such as American writer Richard Louv have argued that nature play is given too little attention because there are only few commercial interests that can make money of it (Talking with the author, 2007). A British organisation named PLAYLINK argues for creating places for play that "...offer children opportunities to: engage with their natural surroundings, be sociable and solitary, create imaginary worlds, test boundaries, construct and alter their surroundings, experience change and continuity, take acceptable levels of risk" (Spiegal, 2007, p. 29). The present play culture is seen as too defensive and risk averse. Too few 'playable spaces' that lack fantasy exist at present.

Organisations such as the Flemish Forest Organisation and the Dutch State Forest Service have taken up the challenge by promoting and developing so-called 'play forests' ('speelbossen'; see Van den Berg, 2000; De Vreese & Van Nevel, 2006; Fig. 6.3). These are special forests or zones within forests meant for play by below 18-year olds and their supervisors (such as teachers, scout leaders). These play forests are inspired by early examples of 'wooded playgrounds' such as those of the Wald-Kinder-Tummelplatz (forest-children-play area; WAKITU) playgrounds in city forests such as the Eilenriede of Hanover (Fachbereich Umwelt und Stadtgrün, 2004). In Flanders, efforts have even resulted in a decree by the Flemish government and a subvention scheme. Special play forest signs can now be found in dedicated forests across Flanders. The Flemish Forest Organisation has drawn up a list of criteria for play forests that deal, for example, with finding appropriate locations, avoiding vulnerable biotopes, ensuring a certain minimum size (2–5 ha), and the forest being situated close to population centres. When establishing play forests, agreement obviously has to be reached with the forest owner and local residents. Moreover, the involvement of the users (i.e. children) should also be ensured during the design and establishment of play forests (Van den Berg, 2000).

Promoting children's outdoor play can have many positive effects, for example related to children's personal development and raising natural awareness, as shown above and also in Chapter 10 'The Forest of Learning'. In The Netherlands, a special holiday camp in the forest has been organised for the children of refugees (Geen psychologisch gedoe, 2006). Children are building huts and play games in nature, whilst focus is taken away from the problems they face in their daily lives.

Differences in city forest use also continue to be influenced by level of income, social status and cultural background. During the earlier 20th century, for example, the more peripheral parks and beaches were used by the poorer part of the population of Helsinki to hang out their washing, clean carpets, but also for children's games, nude bathing and drinking (Clark & Hietala, 2006). In present-day Marseilles, France low income groups and retired people especially use the inner-city public green, while most of the population takes out to the surrounding hills, other more natural areas and local beaches for their recreation (Werquin, 2004). That income and social characteristics play an important role is also described in the previously mentioned study by Gadet (1992). This work shows that the growing group of singles in Amsterdam has its own specific use of urban green space, being more short-term, frequent and dynamic than more 'traditional' uses. Singles use green space in a more socially-oriented way, as a main motivation is to be amongst other people.

Ethnic background is also of influence on preferred city forest use. In the Amsterdamse Bos, for example, picnics are preferred activities of people of Turkish and Moroccan descent (Dekker & Verstrate, 1992, cited by Van den Ham, 1997). Work done by the Dutch research institute Alterra (Buijs et al., 2006) looked at how

Fig. 6.3 City forests are an important environment for children's play. Special 'play forests' can promote outdoor play, which on its turn has positive effects on children's development (Photo by the author)

different groups such as the ethnic minorities and youths evaluated pictures of different landscape types, including forest, heath land, sand dunes and peat meadows. Minorities did not evaluate cultural landscapes such as peat meadows very positively, bluntly stating "I am not a cow". Also in The Netherlands, studies have shown that inner-city green spaces are popular with the main ethnic minority groups; parks are often used for social activities, taking place in larger groups. However, more than half of the minorities questioned did not even know the nature areas around their own city, including the peri-urban forests. Second and third generation immigrants do use more remote recreation areas. Perhaps this is because they have been more in contact with Dutch culture and habits. Moreover, they are often in a better economic situation which makes it possible to spend more time on recreation. The seaside and beach and dune areas are the most popular natural areas outside the city for these minorities. Forests score very low, although preferences of groups do differ. About half of the people of Turkish and Moroccan descent visit forests outside cities. Forests are used more frequently by people originating from Surinam (Jókövi, 2000a, b; Raad voor het Landelijk Gebied, 2002). A study by Burgess (1995) showed that Asian and Afro-Caribbean women in England were hardly using the local woods, because of unfamiliarity – where they live there is often little or no green space and have they have limited experience of recreational woodland – and feelings of fear (see also Chapter 3 'The Forest of Fear').

Finally, men and women are also known to look at forests and landscapes differently. Men tend to take a more 'distant' and overall view, whilst women tend to have more eye for detail (Lehmann, 1999). Women more frequently express fear of using forests and other green areas, especially by themselves (Jorgensen et al., 2006).

Escape from City Rule

In the previous sections, the recreational importance of city forests as a way of relaxing and of escaping from daily routines has been covered. But now we will take the concept of 'escape' one step further, looking at the 'space' and 'frontier feeling' of city forest. How about the long-lasting role of the city forest as being a place on the edge of civilisation, perhaps even beyond the control of the city? A place where activities can be undertaken that do not always tolerate urban daylight, as also discussed in Chapter 3 'The Forest of Fear'?

Thomas (1990) describes the special role that the urban fringe has always had, as a place of conflict, escape, and as interface between city and surrounding countryside. Often, planners have taken a 'zone of problems approach' when looking at the urban fringe. The urban fringe has always been a place were different social groups, different lifestyles (urban and rural) encountered each other, not seldom leading to conflicts. The fringe has also been the place where the greatest demand for recreational land – and for urbanites a place for finding 'rural solace' – has been felt.

Literature is rich in stories of how city forests have harboured outcasts and outlaws, people who did not fit in, and people who voluntarily moved to the edge of urban life. The London commons, for example, were known as a place where 'gypsies and tramps' were hanging out (Reeder, 2006b). The classic stories of Dick Turpin (in places such as Epping Forest) and Robin Hood (linked to Sherwood Forest) symbolise the role of the forest as a shelter for those outside the law (Porteous, 2002; Muir, 2005). In the legends, these characters were despised by rulers who saw them as criminals, whilst local peasants often regarded them as some sort of heroes, standing up to the local lords and their henchmen who continuously diminished people's rights to use forests. Lehmann's (1999) many stories about criminals, terrorists or just simple outcasts that took to the German forests also symbolise the role of city forests as refuge.

But one does not have to be an outcast or outlaw to enjoy the freedom offered by city forests and to appreciate them as areas away from city rules. First of all, the city forest is place for sex. When Lehmann (1999) asked his interviewees about their 'forest experiences', the talk frequently turned to love and sex. However, the author suggests that the forest does not have the same 'sexual importance' anymore, at least not for young people, as formerly it was much more difficult for lovers to find a suitable place.

Geographical connotations still refer to the ancient links between city forests and sex. The Mastbos wood south of the Dutch city of Breda, for example, has a 'Hoerenpad' (Whores' Path). Centuries ago residents of London already complained about 'night-time activities' (read: the undertakings of prostitutes and their customers) in green spaces such as Hyde Park. But during the Victorian age, spirits in London were very puritan, as shown by the prohibition of nude bathing by children and the controlling of the behaviour of courting couples (Reeder, 2006b). More recently, nearby residents and users of city forests such as Scheveningse Bosjes, a wood in the Dutch city of The Hague, have complained about the use of the wood as a sexual meeting place, for example for gay men (Ramdharie, 1995). The woods of Castelfusano near Rome can be filled with cars driven by men, slowly driving past the large number of prostitutes standing along the asphalt roads winding through the forest.

Accounts such as those of the Scottish teenagers (Thompson et al., 2006) show, however, that it is important for cities to offer 'escape opportunities' and to provide places that allow for adventures. The importance of urban green space as 'teenager hangout' is also described by Gallagher (1993). Chapter 8 'The Wild Side of Town' illustrates that some form of 'wilderness experience' close to people's neighbourhoods is much appreciated, for example as source of inspiration and stress relieve. City forests can offer this type of escape. It is important to fully recognise this role, as well as the other – sometimes contradictory – roles as Great Escape. City woods are often relatively small, especially when 'refuge seekers' as well as those looking for 'mass entertainment' – many people are happiest in public settings with other people around, as mentioned by Gallagher (1993) – have to be accommodated. But several successful examples show that even small city forests can fulfil many roles at the same time, helped by the fact that the forest environment makes it easier to zone for different uses and host a large number of people without feeling crowded.

Chapter 7
A Work of Art

Just like the myths described in Chapter 3 'The Spiritual Forest', art is an important part of culture. This chapter looks at the relationships between city forests and art. Woodlands in and near cities have inspired writers, poets, painters and other artists over the ages, as the next section shows. During the 19th century, this artistic interest even led to some of the first citizen movements to protect city forests against urban and other pressures. The second part of this chapter illustrates how city forests can be considered works of art in themselves, providing their owners with prestige and offering attractive environments to recreate. The design and structure of city forests have changed over time according to changing preferences and fashions. In modern times, some city forests have even become part of so-called 'landscape art'.

City forests have also offered attractive settings for art, as they combine a natural environment with the nearness of many people to admire works of art. This is the topic of this chapter's third section. The final part of the chapter shows how art is increasingly being used to strengthen the ties between city forests and local inhabitants, enhancing the joint place making and community building which will also be discussed in Chapter 11 'The Social Forest'.

City Forests as Inspiration

Over time, many writers and artists have found their inspiration in forests. In Germany, for example, forests became associated with a longing ('Sehnsucht') for former times and acted as a symbol of German culture (Lehmann, 1999). This was reflected in many stories, poems, novels, songs and paintings, especially during the 19th century.

City forests have played a special role in inspiring artists. These forests are situated close to the artistic centres that many cities and towns are, and thus comprise a convenient 'inspirational setting' for writers and other artists. Møller (1990) and Kardell (1998) illustrate this inspiring role of city forests in their publications on Jægersborg Dyrehave near Copenhagen, respectively Djurgården in Stockholm. Both authors tell how various famous Danish and Swedish authors and poets

C.C. Konijnendijk, *The Forest and the City: The Cultural Landscape of Urban Woodland*

included the forests in their writings. Through their links with famous writers, poets and artists, these forests also became associated with national culture and later cultural heritage. Møller (1990) describes, for example, how Adam Oehlenschläger, the poet who wrote the Danish national anthem, carved his name in one of the trees in Jægersborg Dyrehave.

Nature, trees and woods have been an important source of inspiration for British writers and poets as well. This is described by Taplin (1989, p. 19), who writes: "like our land, our literature has been wooded". Forests and trees feature prominently in the works of Shakespeare. The Greenwood as a symbol of freedom and retreat is celebrated when Duke Senior takes to the Forest of Arden to find refuge after having been ousted from the court by his brother in 'As you like it' (cf. Taplin, 1989).

City forests and other urban nature areas were among the particular sources of inspiration also in Britain. Hampstead Heath near London inspired many poets, writers and painters, particularly during the 19th century. The poet Leigh Hunt, for example, wrote a series of sonnets titled 'To Hampstead'. Famous writers such as Percy Bysshe Shelley, John Keats and George Gordon Byron visited Hunt's house at the Heath and were also inspired by the artistic appeal of the area. Keats found trees to soothe his troubles and lift his thoughts. Well-known British painters such as John Constable painted Hampstead scenes. The 1863 Watkin Willams song 'Hampstead is the Place to Ruralise' narrated of the newly found recreational importance of the Heath (A history of, 1989; Taplin, 1989).

More recently, J.R.R. Tolkien was not only inspired by Shakespeare's Macbeth when he came up with the idea of walking and talking giant trees called Ents for his Lord of the Rings trilogy. These Ents, like other woods in trees in the work of Tolkien, are a central symbol of life and goodness (Taplin, 1989). It is likely, however, that Tolkien was also influenced by the ancient trees in the landscape of the rural fringes of Birmingham, where he spent part of his youth (Muir, 2005).

Among the city forests that inspired Dutch artists and writers is the Haarlemmerhout. This Haarlem city woodland features prominently in what is considered one of the masterpieces of Dutch literature, the Camera Obscura by Hildebrand (the alias of Nicolaas Beets), which was first published in 1839. One chapter describes, with great eye for detail, the different groups of visitors to this 'walking forest' (Guldemond, 1991; Hildebrand, 1998). The book's main characters and its author are commemorated in the forest by means of a group of sculptures (Bos, 1994). Another Dutch city forest, the Haagse Bos, inspired the Portuguese writer Ramalho Ortigo. He wrote about this the city forest of The Hague in his 19th century book 'A Holanda', a travel account of Holland. According to Ortigo, the forest was the most beautiful of its kind in Europe. Compared to the Haagse Bos, Bois de Boulogne and Hyde Park were only "miserable small gardens" (cited by Anema, 1999, p. 19).

The accounts of writers such as Ortigo and the works of painters provide an important reference – although not a fully reliable one – for city forest history. A 19th-century book by Dutch writer Isaac Anne Nijhoff has been called 'walking with the pen', describing the writer's visits to the forests and parks of Arnhem in The Netherlands. The author writes of an unbroken belt of forests, fields and meadows, primarily owned by a small group of large land owners. The account is a mixture of a well-documented guidebook and a work of literature. British literature has been used as a source for historical study by people such as Raymond Williams. In his book 'The Country and the City', Williams (1973) analyses English literature from the perspective of the changing relationship between city and countryside. Other works of art also help understanding the past of city forests. Dutch painters such as Jan van Goyen painted views of Arnhem's forests and parks during the 17th century (Schulte & Schulte-van Wersch, 2006). The historical account of Belvoir Park near Belfast by Simon (2005) shows a number of 18th century paintings by Jonathan Fischer. Although these paintings are not entirely reliable in their accuracy of depicting the actual situation at the time, they do show some of the main features of city forestry.

Artists and writers were among the first to stress the importance of forests and nature in general as source of inspiration. This was the case, for example, in the ancient Greek and Roman civilisations. Much later, in 19th-century France, a group of artists, poets and writers known as the School of Barbizon had the forest of Fontainebleau as its main object of affection. The School succeeded in creating and popularising an artistic vision of the landscape and the forest. This helped make the forest of Fontainebleau, with its cultural attractions such as the castle and its festivals, become a popular destination for the Parisians during the second part of the 19th century (INRA, 1979; Kalaora, 1981). The School of Barbizon did not only find inspiration in Fontainebleau, however, the group also fought for the conservation of their 'muse'. They successfully protested against plans to partially cut the forest and 'optimise' economic use of the forest that would remain. Partially through a petition and partially by mobilising their fellow Parisians, they succeeded in getting certain parts of the forest protected, including their own favourite spots, such as special tree-rock combinations and remarkable individual trees. Some said: "La forêt de Fontainebleau n'est pas la nature, elle est une oeuvre architecturale" (the forest of Fontainebleau is not nature, it is a work of architecture) (Kalaora, 1981, p. 100).

Writers, poets and painters had a similar impact on the preservation of other city forests across Europe, including the Zoniënwoud near Brussels. This is shown by van der Ben (2000) in his richly illustrated book about the forest, which includes numerous drawings and paintings of forest and landscape scenes. Many city forests had their own 'resident groups' of artists, particularly during the Romantic era. But resident artist groups are still associated with city forests, for example, in various community forest schemes in England, Scotland and Northern Ireland (Konijnendijk, 1998; Simson, 2001a; Simon, 2005).

Art and activism have thus gone hand in hand in the case of city forests. Schama (1995, p. 123) mentions how German painter and sculptor Anselm Kiefer was encouraged in his artistic fight against the mainstream by Joseph Beuys whom he considered "the most creative and aggressively confrontational of Germany's post-modern artist". In the early 1970s, Beuys staged a theatrical (and very successful) demonstration in the Grafenberger Wald outside Dusseldorf. This protest was directed against a proposed conversion of part of the woods into country-club tennis courts. "Together with fifty students and disciples Beuys swept the woods with birch brooms in a kind of ritual exorcism of the bourgeoisie, painting crosses and rings on the threatened trees as if he were affirming the ancient Teutonic religion of wood-spirits" (Schama, 1995, p. 123). Beuys also initiated the project of 'Seven Thousand Oaks' to be planted in the centre of German cities. This project inspired similar activities in countries such as Britain (A. Simson, 2007, personal communication).

Designed Landscapes

We have seen how in France, because of the efforts of the School of Barbizon and other groups and individuals, the forest itself became recognised as a work of art. City forests have been 'works of art' in different ways, because of their natural and cultural-historical values, but also because of their being designed and re-designed according to the fashions of the time. As Rackham (2004, p. 5) writes: "Tree-planting runs in fashions, like architecture or costume". In England, for example, Victorians were fond of the Atlantic cedar (*Cedrus atlantica* (Endl.) Carrière), whilst Georgians preferred using the Cedar of Lebanon (*Cedrus libani* A. Rich). Lawrence (1993, 2006) provides an historical account of how fashions in landscape design influenced urban green space, from French baroque garden and (English) landscape style to post-modern designs. Design of green space, including city forests, has been the result of changing cultures, preferences, fashions and nature views. Whilst, in earlier times, nature was regarded from a rather utilitarian perspective and as something to be controlled, Romanticism, for example, led to a re-appreciation of 'wilder' nature and of trees as prominent symbols of nature, as reflected in art (e.g. Jensen, 1998). Lento (2006) mentions how artists in Helsinki led a movement to re-appreciate nature which also came to influence landscape design. In Germany, the forest aesthetics movement developed as part of Romanticism during the 19th century. In the Eilenriede forest of Hanover, designing for recreation came into focus at the end of the 18th century, for example, in discussions about how to make promenades in the forest. Following the romantic appreciation of nature, the aim became to create 'beautiful forest images' (Fachbereich Umwelt und Stadtgrün, 2004).

The development of the Eilenriede was typical for many city forests. Over time, many 'wild' forests were transformed into hunting parks and subsequently into recreational parks (Rackham, 2004; Muir, 2005). These new hunting and recreational parks were often located in or close to cities. This is described in the case of Tiergarten in Berlin by Hennebo (1979). The area was transformed from an aristocratic hunting domain into a park-like area meant for public recreation. This transformation involved various beautification efforts and the development of bridges, paths and other elements typical for the time. Obviously there was an element of prestige efforts like these. The rulers and aristocracy saw city parks and woodlands as an opportunity to raise their prestige and to show, for example to foreign guests, that they were up to speed with current, international fashions.

Muir (2005) stresses that the reshaping of 'wildwood' into hunting parks and later landscape parks was a gradual process. About the work of leading proponents of the English landscape style such as Repton and Brown, Muir states that they were working with the existing landscape – which often contained woods and open spaces, attractive ancient trees – rather than starting their design from scratch.

The landscape style represents one of the stages in the continuous transformation of city forests such as the Dutch Haarlemmerhout (Brink, 1984; Guldemond, 1991). From the original late-renaissance style (1584–1650), the wood was transformed according to the principles of the formal garden (i.e. baroque) style (until 1780), followed in its turn by the landscape style (until 1890). The Hout was then redesigned in the style of 'gardenesque' garden art, and then – after World War II – according to the rational garden style. Another Dutch city forest, the Alkmaarderhout underwent similar changes (Alkmaarder Houte dupe, 1992). Each of the styles had its own characteristics and 'trademarks'. The baroque style, for example, meant that so-called 'star forests' were introduced to facilitate parforce hunting (a type of hunting using dogs). The English landscape style implied that the former straight lanes inherited from French and Dutch design were turned into organic, winding roads, and that efforts were made to 'imitate' more natural, semi-open landscapes and to incorporate existing natural elements such as large trees. Moving from one style to another meant that many elements of the previous period were destroyed. This raises questions about efforts that focus on one specific style period or date when restoring city forests and parks. As Tuan (2007, p. 195) states, "nostalgia for an idyllic past waxes strong". But do old forest designs meet the demands of modern society? Neither history nor landscape are static and should not be considered as such, although this can cause conflicts with the conservation of cultural-historical values.

In modern times, post-romantic landscapes have been developed where the atmosphere inside does not remind one of the outside environment in any way (Lörzing, 1992). Amusement parks, like the Efteling in The Netherlands, and of course also Disneyworld, are examples of this, but so are zoological gardens, although many of the latter were establishing during the period of Romanticism. Many of these (post-)romantic 'entertainment landscapes' have the character of, or are situated in, a city forest.

In The Netherlands, many new peri-urban woodland and recreation areas were developed during the 20th century to meet the demands of a highly urban population. Blom (2005) shows how woodland design changed and evolved over a time span of a century. The designers of new woodland areas were very much inspired by visits to areas abroad, most notably to England, Germany and Belgium. Wooded areas like the Amsterdamse Bos and the Kralingse Bos in Rotterdam, established during the early 20th century, were still very much based on the English landscape style. The Amsterdamse Bos, for example, looks more like a 'rolling' English landscape than the typical polder landscape of the western part of The Netherlands. That said, the design of the Amsterdamse Bos and similar woods was also influenced by the more modernistic German Volkspark philosophy, which embodied a focus on social use rather than aesthetics in planning and design. The rationality of the modernistic approach is reflected in its inclusion of three (more or less equal) parts: woodland, open areas and water. The Bos was arguably the first post-modern landscape development in Europe (Simson, 2005b).

After World War II, forest planning and design styles in The Netherlands continued to change. New forests such as the Horsterwold in the Flevoland province were adapted to the existing polder landscape, with its straight lines and ditch system. Elements of the agricultural landscape were incorporated into the new forests. Towards the end of the 20th century, woodland design became even more influenced by history. Woodlands such as De Balij – Bieslandse Bos and the Bentwoud near the town of Zoetermeer were seen as parts of complex urban landscapes undergoing continuous changes. Links with historical land use have been recognised and reflected in the design. Parts of the Bentwoud, for example, look like the old manorial estates of previous centuries (Blom, 2005; Van den Berg, 2005).

City forests are valuable from a cultural-historical perspective not only because of their design and history, but also because of specific cultural-historical elements and the links with special events and persons. Koch (1997) explains that forests are bound to harbour many cultural-historical values, as they have often been allowed to develop over a long period of time, where other land uses succeeded each other more rapidly. Also because of increasingly strong forest legislation, the cultural-historical values 'captured' in them could be preserved. In Denmark, for example, registered historical sites and monuments occur approximately three times more often in forest areas than in agricultural areas (Koch, 1997, citing data from the Danish Ministry of Environment).

The cultural-historical values of city forests are not only linked to humans, but also to natural values, as was illustrated earlier by the case of Fontainebleau. Many city forests host ancient and/or exceptional (landmark) trees, for example. An inventory of Belvoir Park Forest near Belfast found 270 single stem trees with a girth of 3 m of more (Simon, 2005). Trees have been considered 'cultural beings', powerful symbolic creatures, symbolising death, eternal life, connection between earth and heaven, justice, and many other things (Clifford, 1994; Perlman, 1994; Jones & Cloke, 2002; see also Chapter 2 'The Spiritual Forest').

Today, the cultural(-historical) values of city forests are often recognised and cherished. The main functions of Fontainebleau, for example, are listed as recreation, nature protection, production and 'artistic' values (Kalaora, 1981; Slabbers et al., 1993). Management of Epping Forest near London has been guided by recreation, natural values and cultural-historical values for more than a century. The forest hosts many historic features, as well as high natural values, the latter especially due to ancient tree population and the unique assemblage of fungi and insects which it supports (Slabbers et al., 1993; The Corporation of London, 1993). The links between nature and culture are also acknowledged in the case of the Wienerwald of Vienna, which is a UNESCO Biosphere Reserve. The reserve is aimed at combining protection of biodiversity with social and economic development, as well as conservation of cultural values (Umweltdachverband für das Biosphärenpark Wienerwald Management & Österreichische UNESCO Kommission, 2006).

Elsewhere, former sacred groves are recognised not only for their former spiritual values, but particularly as cultural-historical legacy and examples of symbolic landscapes (e.g. Arntzen, 2002).

Modern land art also looks at forested and other landscapes as works of art (Fig. 7.1), often developing them from scratch with artistic intentions. Starting from the 1970s, for example, land art ideas and philosophies became expressed through soil, water and plantation in The Netherlands. Large areas of open land were being re-developed. An example of a land art project is the poplar cathedral in the province of Flevoland. In 1987, Lombardy poplars (*Populus nigra* 'italica' L.) were planted in the shape of the cathedral of Reims (Blok, 1994). Another recent example of 'land art' in a forest environment, the Dutch 'Museumbos', is presented in the final section of this chapter.

Fig. 7.1 In the German Ruhr area, abandoned industrial buildings combined with new forests create a unique type of landscape art

Settings for Art

Apart from providing inspiration and being considered 'a work of art' (or land(scape) art) themselves, city forests have also been popular settings for exhibiting art. Cities have always been centres of culture and art, as discussed by Burgers (2000) in his reference to 'exhibited space' as an important use of urban public space. City forests provide convenient, attractive exhibition areas and backdrops for the visual arts, drama and so forth (Fig. 7.2).

City forest history shows how many of these forests have been used as settings for festivals, events, dancing and singing (e.g. Gehrke, 2001). An historical example is that of Riis Skov, a forest owned by the Danish city of Aarhus since 1395. Riis Skov was developed into a so-called 'lystskov' (pleasure forest) with many different attractions, including live music performances and other summer events (Århus Kommune, 2004). In the forests of Aarhus today, artists have been 'hiding' art, such as tree sculptures, so that visitors will find these by surprise. Other modern examples of art in city forests include the tree sculptures and festivals in urban woodland commissioned by the Forest of Belfast partnership in Northern Ireland (Simon, 2005). Wooden sculptures along a 'fairy tale trail' are popular destinations in the Toxovo Forest situated in the St. Petersburg, Russia green belt.

City forests such as Sonsbeek in Arnhem, The Netherlands also host museums, sculpture gardens and art exhibitions. In Sonsbeek, the 'Sonsbeektentoonstelling', a large international sculpture exhibition that includes the work of renown artists such as Henry Moore and Auguste Rodin, has been held every five years since 1949 (Van Otterloo, 2006). Sometimes art in city forests attracts considerable attention from the public and the media, as in the case of Giancarlo Neri's 9 m high sculpture 'The Writer', which was exhibited at London's Hampstead Heath during 2005 (Kennedy, 2005).

In the case of the Masaryk Forest near Brno, Czech Republic, art, forest and links to the past go hand in hand. A trail leads visitors along a series of monuments, set as landmarks inside the hilly, forested landscape. The monuments commemorate famous Czech artists, poets, as well as famous foresters. In the same forest, a special 7 m 'Memorial of Trees' was erected in 1938, with the poem 'Trees' by Czech poet Jaroslav Vrchlický inscribed on it. The monument also has engravings of 'tree-honouring' texts and sayings from across the world (Truhlár, 1997).

The development of new city forests can be connected with art, as in the case of the German Ruhr area. Abandoned factories, coal mines, new forests and art are mixed into fascinating 'industrial art' settings (Fig. 7.1). One example is the 'Zollverein', a former colliery and now World Heritage Site north of the town of Essen. The functional architecture of the site has been developed into a work of art through contributions of leading architects such as Norman Foster and Rem Koolhaas. As nature is gradually reclaiming its territory and woodlands emerge around the buildings, the motto is 'alluring industry in the green'. The industrial woodland sites and the buildings within them offer excellent settings for all kinds of art (Zollverein, 2006, 2007).

Fig. 7.2 Many artists have used city and recreational forests as setting for their work. This sculpture titled 'The Ancient Forester II' by David Kemp is situated in the Grizedale Forest, England (Photo by the author)

Art for Developing Community-Forest Links

Art is thus still very much present in city forestry. Traditionally, art has often been a mere 'bolt-on' addition to city forestry and other projects in the public realm, often dropped in at the last minute, for example, to fulfil a planning condition (Simson, 2001a; Trodden, 2007). However, it has become recognised and accepted that art has potentially a far greater role to play than mearly as a decorator of space, and none more so in city forestry projects (Simson, 2001a).

During recent years, the role of art in making meaningful places, creating identity and providing orientation in the vastness of cities has become stressed. Perhaps it is better to speak of 're-making' places. Art often helps to re-define the identities and community ties of place, as much as it can help with turning space into place. Art projects can help develop and re-develop place identity and attachment. But in order for this to materialise, art has to be regarded as an integral part of the public realm and people's experience of the city (Trodden, 2007).

In city forestry, art can help in enhancing linkages between forests, cities and communities (Fig. 7.3). Davies and Vaughan (2001) see art as tool to generate widespread citizen support and to create a culture of 'ownership' by local community, an approach which is actively applied in the case of England's Community Forests. Citizens can be engaged, learn about the forest and even leave their own

Fig. 7.3 Theatre performances as shown here in the Sonsbeek woodland park of Arnhem, The Netherlands, can help build stronger ties between local residents and city forests (Photo by Janette Derksen)

creative marks. Clifford (1994) agrees that the involvement of artists, as well as craftspeople, and the use of performances, festivals, dances, storytelling and poetry can help establish stronger links between a city forest and the local urban community. Simson (2001a, p. 29) states: "On the emotional, intuitive or psychological level, the role of art may be to stir, to cause wonderment, amusement, contemplation or surprise." In the longer term, art can supply and reinforce sense of identity and loyalty to a place, such as a city forest. A sense of belonging can be developed amongst individuals as well as communities.

Trodden (2007) mentions that art should be considered during the different stages of planning and designing processes, for example when they are consultative in nature. Focusing on visual art, Simson (2001a) provides several ways of involving artists in urban forestry with the aim to develop the sense of ownership and pride amongst local communities. An artist can be selected to do the work independently, or alternatively, a collaborative process between artist and local community can be developed. Artists can be asked to 'pull out' ideas, concepts, fears and aspirations from the community and reflect these in their art. In this way, art can be used as a true 'gateway' to community awareness about the local forest. Art's often controversial character should be considered, although this could also help evoke interest and emotions.

In this context, the role of the artist is that of a mediator and enabler. In fact, this has always been an important role of the artist, who had to 'bridge' the rift between different views of nature and forests. This was especially the case from the moment when more utilitarian views of nature started to conflict with those that idealised nature, in particular during the Industrial Era. Taplin (1989) illustrates this as he describes an increasing disquiet amongst English writers with the progressing overexploitation of nature, the 'destruction of trees' and the disappearing of the traditional rural landscape (cf. Williams, 1973). Writers and poets shared a reverence for nature, which they saw as vital to health, spirituality and imagination. They also informed people about the threats to nature without making the situation seem desperate. Or, in the words of Taplin (1989, p. 19): "Only art, and perhaps poetry in particular, can bring terrible things home to us without causing despair".

Different, often opposing nature views are not easily bridged. However, in the post-industrial era, where land is short and demand for it intense, different views and land uses have to be combined. Professionals such as foresters and landscape architects try to manage this, but artists can be important enablers, assisting in re-engaging people with their city woodland places. Their work can help show that utilitarian, romantic and other views of forests all have their place and do not necessarily exclude each other.

A recent example of artists working as enablers is the project 'Homelife' at Chigwell Row Wood Local Nature Reserve (Green Places, 2006). The project encompasses an artist-led collaboration with the local community to explore the local habitat and its surroundings. The area includes an ancient woodland which is frequently abused by people from the nearby housing estate, for example for fly tipping. Not surprisingly, the community has a rather ambivalent attitude towards the woodland. During spring and summer 2007, with the involvement of an organisation called Epping Forest Art, artists acted as mediators in building better relationships between people, and between people and 'place'. Activities included site-specific dance, the making of music videos, the of recording mobile phone tunes with sounds from the forest and installing of a live web cam that showed what happened in the wood at different times of day.

As mentioned, the English Community Forest programme has also been very active in involving artists and art in their various activities to create wooded landscapes around large agglomerations. Developing better links between local people and local environment is the main objective (Davies & Vaughan, 2001; Simson, 2001a). Other partnerships throughout Great Britain and Ireland have used art in a similar way. Elsewhere, the Flemish Forest Organisation, a Belgian environmental NGO that promotes, among other issues, the establishment of new city forests actively employs arts and art events to create better links between local people and their forests. Each summer, they organise a series of 'Movies in the Forest', for example.

Place making, or rather 're-making' place is a major challenge when new city forests are created. In the Ruhr area, art plays an important role in making people aware of the changes in their local environment, linking the industrial heritage to modern woodland. This can help overcoming people's negative feelings about the loss of what once was and restore pride in the community (H.-P. Noll, 2007, personal communication). Art projects such as those at the Zollverein complex mentioned above aim to develop the industry-new woodland area into a recreation and leisure area for the entire family (Zollverein, 2006).

In The Netherlands, artists have been asked to contribute to the woodland design process, rather than develop isolated land-art type projects. Design workshops have been held with forest planners, landscape architects and artists. In one case, the participating artist involved children of neighbouring schools, asking them to draw their 'perfect' forest and trees. A participating French artists developed the concept of the city forest as a supermarket, where the different recreational functions all have their own particular, recognisable place (Van den Berg, 2005).

An interesting example of how city forest design and art can go together is that of the so-called 'Museumbos' (Museum Forest) project near the Dutch city of Almere. In 2001, a private foundation, together with the Dutch State Forest Service, set up a competition that asked artists and architects to design 'forest rooms' of 100 by 100 m. Artists were asked to apply a personal, original approach to the concept of 'forest' in their projects. The best projects would be awarded by actually being realised within the Almere city forests. More than 100 project proposals were submitted, out of which eight were selected for realisation (and so far four projects have been implemented). The winning 'forest room', titled 'Tempo', proposed to plant one lime tree (*Tilia* sp.) within a grid each for, for 100 years. After 100 years, the oldest tree can be replaced. In 2005, the first tree was planted by schoolchildren from Almere, who also buried a bottle with 'surprises' at the base of the tree, to be dug up when this tree will be removed in the future (Baaij, 2002; Museumbos, 2007).

This final example brings us back to the historical links between art and city forests. Not only do city forests provide inspiration and a setting for art, as described in this chapter. Artists also give something back to city forestry, as they can assist in place making and place re-making, making sure that urban people are not moving further away from forest and nature. This mutually beneficial relationship is essential to the city forest as a Work of Art.

Chapter 8
The Wild Side of Town

In 1986, the book "The Wild Side of Town" authored by British broadcaster and environmentalist Chris Baines was published. The book coincided with a popular BBC television series and had a major impact on people's appreciation of urban wildlife in Britain. Baines wrote: "The green space in towns and cities is much more than a simple, slightly degraded duplicate of pre-war farmland. For a great many species, the town is a much better place to live in than the countryside ever was." (Baines, 1986, p. 29). Various studies have confirmed that urban areas often harbour a perhaps surprisingly high variety of species of plants and animals (e.g. Cornelis & Hermy, 2004; Gustavsson et al., 2005; Alvey, 2006; Lorusso et al., 2007). A recent magazine article (Theil, 2006) confirms that there is a place for nature even in the most urbanised area. It refers to research by Munich's Technical University which found more species and more diverse habitats in selected big cities than in any national park of nature reserve. Berlin is home to two-thirds of all bird species in Germany, while Zurich hosts ten times more foxes, hedgehogs and badgers per square kilometre than the surrounding countryside. Part of the explanation is that cities offer a mosaic of habitats and microclimates, from pond-filled gardens to industrial brownfield sites and, of course, city forests. Moreover, urban green spaces harbour a large number of exotic species. A study of South-African towns, for example, found two-thirds of all woody plant species to be alien species (e.g. McConnachie et al., 2008).

City woods of different sizes play an important role amongst the wildlife habitats in and near cities. But perhaps more importantly, to many city dwellers they also offer 'real nature' close to where they live and work.

But, how is 'nature' defined? Moreover, what about related terms such as wilderness and 'wildness'? As this chapter will show, the meaning of nature, wilderness and other terms has changed over time, in line with changing nature views. Yet, nature has often been seen as 'the other', as that what is non-human and exists without human interference. Wilderness used to be associated with the deserted, savage, desolate, the barren, with places on the margins of civilisation "where it is all too easy to loose oneself in moral confusion and despair" (Cronon, 1996b, p. 70). This negative notion had changed drastically, however, by the end of the 19th century, when during the period of Romanticism, wilderness became sacred and associated with the deepest core values of the culture that created and idealised it.

The changing concept and role of 'wilderness' is discussed in greater detail later in this chapter on the role of city forests as 'wild side of town'. First, the role of city forests in the emergence of the nature conservation movement is presented. Then city forests will be considered from the perspective of changing nature views, as nature is in the eye of the beholder. Expert definitions of what is nature, what is wild or wilderness will not suffice, particularly not in an urban environment. What matters is what different urban dwellers perceive and experience as nature and wilderness. Knowledge about this can help planners and managers in developing city forests as a true wild side of town, as a much needed antidote to urban environments of asphalt, steel and concrete. This is the topic of the final part of this chapter.

City Forests and Nature Conservation

City forests across Europe have featured prominently in the development of the relations between humans and nature. Initially, they harboured wild and dangerous nature, as Chapter 3 'The Forest of Fear' describes in greater detail. Between 1640 and 1650, for example, wolves were still considered a plague at the Forstenrieder Park near Munich (Ammer et al., 1999). Although forests and nature areas were gradually pushed away from cities and towns, wild animals returned after periods of war, epidemics and subsequent population decline, as in the case in Germany, where wolves and bears had a revival following the Thirty-Year War (Lieckfeld, 2006).

During the main part of their history, city forests were primarily used for extracting timber and other products, as Chapter 4 'The Fruitful Forest' has shown. Biodiversity was important in terms of the range of products forests could provide, as well as for hunting purposes. The aristocracy even imported game from elsewhere to their hunting forests for the purpose of enriching the experience of the chase – and of their taste buds.

During the Renaissance, there was some space for 'wilder' nature in and near cities. More rural or natural areas adjoined the gardens of some Italian villas. These areas sometimes included a wood (called 'bosco') where nature and natural processes were given more freedom. As Lawrence (2006, p. 272) writes: "Only in the *bosco* outside the walls of the garden proper or along roadways were trees allowed to grow closer to a full size".

As described in previous chapters, when cities developed, in particular during the Industrialisation but also earlier, many surrounding forests became a victim of urban expansion and the need for more agricultural lands. The city became a locus of power that operated in the natural world, sweeping everything towards its centre (Merchant, 1996). In the case of Stockholm, Nilsson (2006a, p. 105) writes: "Old forests, natural green areas and open spaces were successively taken over by houses, streets, roads, industrial enterprises, shopping centres and schools, but also by parks and gardens, churchyards, cemeteries, allotment gardens, sports grounds and so on." 'Wilder' nature was being replaced by more 'cultural' and controlled green space.

Partly in response to the destruction of nearby nature and forests, a re-appreciation of nature emerged. This movement was initiated before the Industrial Era during the Enlightenment, but really gained strength during the early stages of industrialisation and the period of Romanticism. Nature became (re-)appreciated as antidote to 'corrupt' city life, primarily by the upper classes of society. During the 18th century, for example, a countermovement emerged against the more formal style of landscape design (Hennebo & Hoffmann, 1963). Nature became the new garden idea, and the "Garten war gleich Natur" (the garden was the same as nature), as described by Hennebo and Hoffmann (1963, p. 19). Garden and landscape architects such as Brown, Repton and Lenné designed according to the (English) landscape style, based on sound understanding of natural processes. Tree groups and woods, sometimes relicts of original, natural landscapes, were essential to the landscape style.

Frederick Law Olmsted was one of the landscape architects that focused on 'designing with nature' (Spirn, 1996). He introduced the 'wild' appearance to urban parks in the USA, with the motivation to bring the advantages of natural scenery found in places such as Yosemite to those who could not travel that far. Nature was thus brought to people's doorsteps, for free use and enjoyment of all. This is not to say that Olmsted let nature do the job. His parks and other projects were highly designed, combining the artificial with the natural and introducing exotic plant species along with native ones. Olmsted found it appropriate to apply the same management to trees in urban parks as to those in rural woodlands. He also mentioned that a landscape was never finished after its construction.

Trees and woods were important to Olmsted. As important icons of both terrestrial nature conservation and nature destruction, trees and woodland also played an important role in the re-appreciation of nature described above (Jones & Cloke, 2002). Irniger (1991) mentions how descriptions of the falling of trees in stories and poems came to symbolise the 'rape of nature' as early as at the end of the 18th century (cf. Taplin, 1989).

The nature conservation movement matured during the second half of the 19th century. City forests were often of particular concern to the first nature conservationists. Fontainebleau near Paris, Jægersborg Dyrehave north of Copenhagen and the Zoniënwoud south of Brussels were amongst the first natural objects to be protected successfully. These city forests were a major 'cause célèbre' for the nature conservation movement (e.g. Olwig, 1996; Konijnendijk, 1999). Artists, writers and later also naturalists and landscape architects played an important part in the re-appreciation of nature. The case of Fontainebleau south of Paris, described in Chapter 7 'A Work of Art', is especially well known. Together with other artists and members of the bourgeoisie, a group of painters called the School of Barbizon successfully strived to get the forest protected against overcutting and other assaults. As a result, (part of) Fontainebleau became the first nature reserve in France (Kalaora, 1981). But also elsewhere did writers, painters and other artists play a leading role in urban nature conservation. In Helsinki, for example, they led a 1920s movement of urban romanticism and idealising of nature (Lento, 2006). Artists were supported by architects, biologists, and landscape architects such as the American Olmsted 'brothers' (actually the son of Frederick Law Olmsted and his step-brother) who advocated for the creation of a forest park near Portland, Oregon, USA. They argued that future generations were likely to appreciate the wild beauty and grandeur of the tall fir trees in the local forest (Houle, 1987).

From these initiatives of artists, professionals and influential individuals, many nature conservation societies emerged throughout Europe and other parts of the world. In Britain, for example, the Hampstead Heath Protection Society was established in 1897 to help maintain the 'wild' character of this nature area near London (A history of, 1989).

Designing of green spaces 'together with nature' is not a new phenomenon, as discussed earlier in the case of Repton, Brown and other architects that implemented the landscape style.

City authorities gradually followed suit. The early 20th century 'Stockholm Style' in urban planning (see also Chapter 10 'The Forest of Learning') included the preservation of nature for recreation as well as focus on the experience of nature by city dwellers (Nilsson, 2006b). The Amsterdamse Bos was established starting from the 1930s with a 'natural' Dutch forest in mind, resulting in a species composition with 95% broadleaves (Balk, 1979).

In particular after World War II, local and even national governments got involved in conserving and promoting urban nature. Remaining natural areas in and near cities, such as ancient woodlands, gradually became protected. This was the case for Highgate Wood of London, a 28 ha ancient woodland. It was appointed as Site of Metropolitan Importance for Nature Conservation in 1999. This decision was a direct result of the concerns of organisations such as the London Natural History Society, who objected to the ongoing efforts to the managing Corporation of London to treat the wood as a park, for example by introducing ornamental tree species, asphalting roads and removing dead wood (City of London Corporation, 2007). This example clearly illustrates the emerging tensions between different interpretations of nature, from cultivated and tame to more 'natural' and wild.

During recent decades, nature and the environment have moved up on the political agenda. Concepts such as sustainable urban development have come to influence urban nature policies, as well as city forestry. The United Nations Conference on Environment and Development (UNCED) in 1992 was influential in promoting the role of urban nature, as well as in involving local communities in taking care of their own environment. Related processes such as the Ministerial Conference on the Protection of Forests in Europe have helped to 'translate' the key ideas of UNCED into forestry (e.g. Krott, 2005). Sustainable forest management (SFM) has become a key concept, as it considers the social, cultural, environmental as well as economic dimension of forestry. City forestry has generally not had much difficulty in implementing this concept, as it has been based on principles of multiple use for a very long time (Otto, 1998; Konijnendijk, 1999; Krott, 2004). However, one change has been a shift towards 'closer-to-nature forestry', which attempts to leave more room for natural processes – such as natural decay of dead wood, natural regeneration – in forest management (e.g. Konijnendijk, 1999, 2003; Berliner Forsten, 2001).

A good example of how natural values are promoted in modern city forest management is the Amsterdamse Bos in The Netherlands. The 'Bos' (forest), formerly known as 'Boschplan', was initially developed as a recreation area with balanced parts woodland, open area and water, and with active recreation in mind, much in the tradition of the German Volksparks. Rather unique at the time was that only native tree species were used for establishment (Balk, 1979). According to a policy plan developed during the 1990s, the Bos was divided into four zones of differing naturalness. In the park forest zone, recreation and other human activities are favoured. Forest management will always be necessary in this zone, especially in order to maintain the special landscape structure of the space. The same goes for the forest border zone. In the nature forest zone the focus is on nature for nature leisure seekers. Here, forest management measures will be limited to a minimum. In the nature zone, nature is favoured over leisure seekers. This zone concerns ancient cultural landscapes that always require considerable management. The last zone is the urban fringe zone, where human use – again – is important. Management in this zone does not really differ from that of the park forest zone (e.g. Van den Ham, 1997). These plans were developed further in a new concept policy plan for the period 2000–2010 (Dienst Amsterdam Beheer, 2002).

Conservation and development of natural values have thus gained more prominence in city forestry. This is also demonstrated by the number of cities have national parks and nature reserves within their boundaries or nearby (Konijnendijk, 1999). The heavily forested National Park Losiny Ostrov in Moscow, for example, covers 128 km^2, one quarter of which is situated within the city. This means that 10% of the Russian capital is covered by this national park (Sapochkin et al., 2004).

The links between cities and nature are not only the story of nature preservation. During recent years, efforts have also been undertaken to bring nature, and in particular forests, back to the city. Many European countries have started to afforest abandoned agricultural and industrial lands in and near cities. Denmark, for example, has an ambition to double its forest cover by the middle of the 21st century. In the Danish afforestation programme, areas near cities are prioritised, as these are believed to generate the greatest benefits in terms of recreational opportunities, drinking water protection, and so forth (Kirkebæk, 2002). Other sparsely forested countries such as Belgium (and in particular its Flanders region), Britain, Ireland and The Netherlands have very similar afforestation programmes (e.g. Van Herzele, 2005; Nielsen & Jensen, 2007). The latter country has been a front runner in the rise of the so-called 'nature development movement' starting from the late 1970s. As a reaction on an industrialised and urbanised society, increasing threats to the environment and greater societal appreciation of 'wild' nature, projects were started to 'give back' land to nature. Agricultural areas have been taken out of production and river banks given back to the rhythm of nature. Conditions have been created for spontaneous development of vegetation and ecosystems. Where needed, nature is 'helped', for example by reintroducing large herbivores (Fig. 8.1). 'New nature' thus emerged, even in and near cities (e.g. Blok, 1994; Metz, 1998; Koole & Van den Berg, 2004). That human interference can promote biodiversity was acknowledged

Fig. 8.1 Natural processes are an important driving force in the emergence of new woodland landscapes in the German Ruhr area (Photo by the author)

in a study of the Dutch Alkmaarderhout city forest (Alkmaarder Houte dupe, 1992). Human interference has resulted in a rather diverse forest, with a larger-than-usual share of old trees – as urban dwellers object to tree felling and regeneration is lagging behind – and many exotic species, for example plants from people's neighbouring gardens.

In some cases, new nature has also developed without much human interference, as in the case of some of the abandoned industrial brownfield sites of the German Ruhrgebiet (Ruhr area). Spontaneous woodland development on these sites has resulted in a new type of 'wild woodland' that is comparable to natural or semi-natural habitats and often harbours endangered species (Kowarik & Körner, 2005; Keil et al., 2007; Fig. 8.2).

Nature conservation and nature development in and near cities, for example in the form of city forest conservation and development, can also have some drawbacks. Chapter 3 'The Forest of Fear' already introduced some of the negative aspects associated with city forests and nature, such as peri-urban forest fires. At a recent Forestry Debate Day in The Netherlands, some of the negative health effects associated with bringing (wilder) forests and nature back into the cities were discussed. Mosquitoes, ticks and foxes, for example, act as vectors for various diseases, some of which are very serious (Den Ouden & De Baaij, 2005). Sometimes city authorities have to turn to radical actions to make sure that 'wild nature' does

Fig. 8.2 In countries such as The Netherlands, large herbivores have been introduced to many city forests to help create more natural landscapes (Photo by the author)

not run out of control. The Canberra city government, for example, started feeding birth control pills to the city's abundant kangaroo population (Australia eyes pill, 2006). After a five-year period of drought, many of these animals had turned to the city to find subsistence, thus causing major damages.

Nature Is in the Eye of the Beholder

A recent news story (Mass hysteria after, 2006) describes how a snake was killed in a Nepalese school situated in a village outside of the country's capital Kathmandu. The result of this was that many of the schoolchildren present fainted, as the snake is considered holy by many Hindus. In the end, the head master of the school had to apologise for the killing. In a case in The Netherlands, the Dutch State Forest Service was accused of animal neglect by animal rights groups. In an attempt to make their areas more natural, the forest service has been using free-roaming Highland and other cattle, as well as red deer, for example in the Oostvaardersplassen nature reserve near the town of Lelystad. However, when they did initially not interfere when many of these animals died during the harsh winter of 2004 – as interference would be 'unnatural' – the conflict with animal rights groups emerged (Teletekst, 1999; Vera, 2005).

These examples shows that people have very different perceptions and preferences when it comes to nature. In the 'grazers' case of The Netherlands, Vera (2005) used international research to demonstrate that mortality rates among free-roaming cattle and red deer populations in the Oostvaardersplassen reserve had not been unusual for a harsh winter. Still, this was a type of 'naturalness' which many people did not seem to appreciate. Their overall reactions to the introduction of large cattle had been primarily positive (Maasland & Reisinger, 2007), but the consequences of treating these as part of the ecosystem had not experienced a similar reaction.

While one person may prefer the 'orderly' nature of a well-maintained private garden, someone else turns to the wild wood outside the town. Appreciation of nature is dependent on personal characteristics as well as on cultural background. 'Natural' has different shades of meaning in different cultures and at different times. Nature features in the physical realm as well as realm of cultural ideas and norms (Olwig, 1996).

Jette Hansen-Møller (cited by Oustrup, 2007) has defined nature view as the jointly understood or held meanings that a group of people agrees on and that are expressed through the shaping of nature and landscape, but that are difficult to put in words by the individual. Nature view will thus mirror, for example, the cultural community people are part of. Nature view changes with culture and with changes in society, as shown in the previous section. It influences the way in which people experience, among other, forests. Forest experience can be emotional (the forest as a refuge, source a renewal, with focus on sensory aspects), physical (offering opportunities for unfolding) and symbolic (relating to values and meanings given to the forest).

Cronon (1996a) stresses the importance of a more historical and cultural way of understanding nature. There are many contradictory meanings of nature in the modern and post-modern world. Different 'natures' exist and none of these are 'natural' as such, as they all are cultural constructions that reflect human judgements, human values and human choices. In other words: what is nature is defined by people. This leads to criticism of nature as a 'naïve reality', of nature without any cultural context. This nature as a moral imperative, as something intrinsically 'good' in a corrupted society, is a product of European Enlightenment and lies at the basis of the nature conservation movement. Nature is thus seen as an Eden, as original nature lost through some culpable human act that results in environmental degradation and moral jeopardy (cf. Kockelkoren, 1997; see also Chapter 2 'The Spiritual Forest'). This has encouraged mankind to try to 'reinstate' nature, to celebrate particular landscapes as 'gardens of the world', to look for a 'perfect' landscape, a place so benign and beautiful and good that it is imperative to be preserved.

As Cronon (1996a) makes clear, the search for these 'new Edens' results in conflicts over particular visions of nature. Where one group of city dwellers wants to plants trees to create a neighbourhood woodland, another group objects to this when the trees block their private view. Conflicts over contested nature can be violent when both sides defend their version of Eden (see also Chapter 5 'The Forest of Power' and Chapter 12 'A Forest of Conflict'). Korthals (1998) mentions that the 'battle' over nature has intensified and that nature has become equal to disagreement. The cognitive, ethical and aesthetic 'valuation perspectives' of nature held by different groups and individuals vary widely in modern Western society, where there no longer are certain fixed, leading principles. People's respective views of nature are related to their identity, to who they think they are and the kinds of lives they want to live. 'Ideal' nature often becomes constructed, taking extreme forms that are highly 'artificial', as is the case in many city parks, zoos, but even more so in theme parks such as Disneyland. Idealised versions of nature become part of consumerism, of a consumable experience. Nature becomes a commodity.

On the other hand, there is still the radical 'otherness' of nature (Perlman, 1994; Cronon, 1996a, c). This otherness lies forever beyond the borders of our linguistic universe; we are unable to define it. Merchant (1996) uses the term non-human nature, to be recognised as an autonomous actor. This perspective is also taken by Jones and Cloke (2002), who attribute agency to trees, recognising them as actors of their own, for example in place making (cf. Rackham, 2004). In geography and other fields, there is thus a need to recognise and accommodate the presence – and active role – of non-human actors. Nature is not merely inscribed upon by human culture and practice, but it is 'pushing back', offering creative inputs. In a particular place, relationships between the human and non-human community present will have to be considered and balanced, in recognition of connections to the larger world through economic and ecological exchanges. Nature can also be seen as a demonic other, as an uncontrollable avenging angel (Cronon, 1996a, b) that brings about hurricanes, floods, wildfires. But the demonic other also manifests itself in the deadly attack of a mountain lion or bear on a jogger and even in the disease transmitted by a forest tick (see Chapter 3 'The Forest of Fear').

That different people and different cultures have highly varying views of nature is illustrated by Proctor (1996). The author cites a 1993 study by Kevin Lynch who investigated Caribbean Latinos in New York City, USA. These city dwellers had very different and much more utilitarian views of city nature than other New Yorkers and, for example, used littered median strips to grow vegetables. A nature experience for them was catching bluefish from a party boat. Results from a study among residents and green space managers in Christchurch, New Zealand, confirm that 'wild' and 'nature' had different, confusing meanings. Vegetation in the city was considered at least as much in social and cultural terms as in ecological terms. The issue of exotic versus native vegetation was hardly raised by those involved in the study (Kilvington & Wilkinson, 1999), although it has been frequently debated, also in the context of city forests (Gustavsson et al., 2005). That 'real nature' does not have to have much in common with natural vegetation is also shown by the example of the Ayazmo Park in Stara Zagora, Bulgaria. An unpublished survey among residents showed that this man-made forest park, which harbours many exotic tree species, is seen as a 'typical' example of wild, Bulgarian forest (F. Salbitano, 2002, personal communication). Studies in Germany have indicated that Germans (and walkers in particular) appreciate a certain level of wildness, but also that this needs to be balanced with an impression of a 'clean' economic and recreation forest. Wildness is good, as long as it is clearly controlled (Lehmann, 1999).

Also when speaking of city forests, the question thus has to be asked: whose nature are we talking about? Different nature views need to be recognised and anticipated if city forestry is to be successful. Our definition of different 'natures' will inevitably be anthropogenic, but – as Proctor (1996) also stresses – this is not the same as anthropocentric. The author refers to the work of Michael Pollan who argues for a garden ethic in our modern-day relation with nature. A gardener, Pollan (1991) states, has a legitimate quarrel with nature, tends not to be overly romantic and does not take for granted that man's impact on nature will always be negative. This thus contrasts with the view that man has destroyed Eden. Rather, man can form an alliance with nature in search of 'better' landscapes.

The question of 'whose nature' also has had an important link to issues of equity. Over time, the dominating views of nature have been expressed by the (urban) elite and urban green spaces – including city forests – were created and managed according to these views. A result of this are today's city forests with the basic design of hunting domains or of prestigious, grand recreation areas with exotic trees. During the 19th century, people like Frederick Law Olmsted objected to the dominance of the elite in using and shaping nature, for example, in debates about Yosemite National Park. Olmsted used Britain as a bad example, as he considered the most preferred and unique natural scenes there to be a monopoly of very few, very rich people (Spirn, 1996).

White (1996) describes another conflict in different people's nature view, namely that between work and play. Environmentalists have often seen work in nature as destruction, perhaps apart from certain kinds of archaic work, typically small-scale farming by peasants that demonstrates close knowledge of and connections with nature. But work also means knowing nature, establishing attachment to physical nature and creating knowledge, as reflected in the gardener perspective presented by Pollan (1991). Working and living with forests provides good opportunities for following the forest's rhythm and seasonal variation and enables closer ties to be established between man and nature (Oustrup, 2007).

As said, the question of 'whose nature' is important when speaking of city forests. In the case of countries such as Belgium (Flanders), Britain, Denmark, Germany, Iceland, Ireland, Italy and The Netherlands, for example, answers to this question influence the ongoing establishment of new forests close to cities. Who defines what these forests will look like? Who prefers forests over the open, mostly cultural landscape that they replace? Some people will argue that the decision lies with foresters, landscape architects and other experts. Others will be of the opinion that the decision needs to be taken by local inhabitants. But of course there is no such thing as the 'typical' local inhabitant. One will thus have to recognise and deal with a variety of nature views.

Searching for the Urban Wilderness

What do people look for in city forests when they want to find 'real' nature, however defined by them? As seen, people's nature views are very different. Still, there are some similarities. Lehmann (1999) found in his interviews of German forest users that mysterious qualities are often ascribed to forests. Forests are dark and dense. People mostly avoid night and darkness, walkers and visitors return home before dusk. There is a fear to get lost, but on the other hand, a 'real' forest needs a certain minimum size. Mornings with their nice temperatures and bird song are appreciated. The time around noon, when things get silent, is often described (also by people working in the forest) as a less pleasant time. Forests represent the different seasons, although seasonality is less important than in the past. Seasonality does matter, however, to mushroom and berry collectors, to beekeepers, to hunters. A study by Macnaghten and Urry (2000) looked at what people perceived as key characteristics of a forest. Replies referred to opportunities to get away from daily life, to disconnect from a stressful job, and to experience nature on one's own body. Forests were seen as the 'right' place for people to experience colours, seasons, forms, rhythms and sounds (cf. Oustrup, 2007 as presented in Chapter 1). Forest experiences were regarded to refer to myths, literature and poems. Experiences were seen as relating to certain childhood memories. Jorgensen and Anthopoulou (2007) found in a study of elderly forest users in the United Kingdom that woods were particularly valued for their links with the past and the opportunities they offered for immersion in the natural world. A study among inhabitants of Wales by Henwood and Pidgeon (2001) showed that forests and trees symbolise nature that will always be there, in an over-urbanising landscape.

City forests are often seen as the closest one can get to 'wilderness' inside or near a city, even though 'forest' refers to a more domesticated form of nature. City forests provide robust, larger-scale nature, where one can get lost in more than one sense. But what is 'wilderness'? Is it not contradictory to have wilderness in the city? As mentioned earlier, Cronon (1996b) discusses the development of the concept of wilderness and stresses that it is a product of civilisations rather than an area without contamination of it. Rooted in the period of Romanticism, and on the initiative of the elite, wilderness has become sacred and associated with the deepest core values of the culture that created and idealised it. In North America, for example, wilderness was first of all the sublime, with the supernatural laying just beneath the surface. The sublime evokes strong, often overwhelming emotions that are not always very pleasant, although the sublime has gradually become tamed and domesticated, for example in national parks. Wilderness to North Americans is also the (vanishing) frontier, a last bastion of rugged individualism, representing freer and more natural wild lands. Today, 'real' wilderness is sought today in travels to places such as Amazonia or Antarctica. Slater (1996) speaks of people's hope, from a feeling of deep nostalgia, for rediscovery of paradise, a return to nature, but also for the construction of a new Eden. From an area with lack of utility, uncultivated and uninhabited by human beings, wilderness has gradually become an invitation to adventure, to incipient nostalgia for unspoilt origins.

Interesting is also the distinction between concepts such as wilderness and jungle (e.g. Slater, 1996). Jungle derives from Sanskrit ('jangala'), where it means 'dry' or 'desert', as well as dwelling-place of wild beasts (i.e. much like wilderness). But unlike wilderness the term also suggests tortuous complexity, getting lost, danger; it has a much more negative connotation than wilderness and does not stand for an area of deep reflection like wilderness. This is confirmed by the use of 'urban jungle' as a negative term, being associated with getting lost, danger, and the like (Lehmann, 1999).

By definition, wilderness is situated on the fringe of civilisation. The central paradox is that wilderness embodies a dualistic vision in which the human is entirely outside the natural (Cronon, 1996b). So our very presence in nature represents its fall. The reality today is, however, that wilderness is a very different and much broader concept, referring even to areas that are situated in urban areas, close to our homes, but that seem to have levels of human interference to contrast with the heavily controlled and artificial environments that dominate our cities (e.g. Kowarik, 2005).

It is clear that there is a place for both 'wild' and more controlled nature in cities. A literature review by Jorgensen et al. (2006) shows that people need both managed and 'wilder' areas in these living and working places. They tend to prefer more managed landscapes close to their house, but also need 'wilder' green areas, including woodlands, close to their neighbourhood. Kaplan (2001, cited by Jorgensen and Anthopoulou, 2007) states that tended nature is linked with 'neighbourhood satisfaction', whereas wilder nature is connected to 'nature satisfaction'. The author also suggest that well-tended nature represents community place identity; wilder areas may foster individual place attachment. In other words: the neighbourhood park is more for community purposes, while the wilder parts of the city forest are directed towards the individual and its relation with nature.

Although concepts and perceptions of 'wilderness' differ, it is clear that there is a link between city forests and a 'wilder' form of nature. As described in Chapter 1 'Introduction', city forests are 'space', i.e. a less controlled, less familiar area where new and adventurous experiences are possible. The tension between the role of city forests as space, as well as familiar place is illustrated by a study in Finland which showed that people prefer the idea that a forest is nature, but still they want management as well (Tyrväinen et al., 2003). City forests often harbour high levels of (recreational) facilities, which can also conflict with giving a 'wilderness impression'. Managers of Epping Forest have therefore made the choice to keep the number of information signs and panels to a minimum. In a study by Lipsanen (2006, pp. 245–246) of appreciation of a forested area near Helsinki, one conclusion was that "one needs to take into account the natural morphology of the area and build trails, shelters and other elements in a way that reinforces the existing natural forms and does not detract from them".

As wilderness often holds associations with 'unspoilt' nature and absence of human interference, perhaps it is better to speak of 'wildness'. City forests that are clearly managed can still be 'wild', leaving room for natural processes, dead wood, natural regeneration, and so forth.

Working with Nature

It can be expected that the demand for parks, wilderness and other protected areas will increase as a result of people's growing appreciation for nature. The call for a better quality of life and environment in urbanised societies will enforce this development. This chapter has shown that city forests are appreciated as 'wild side of town', even though what is meant with 'wild' is not entirely clear and different on context and perspective.

There is a clear link between wild(er)ness, nature and urban areas, although this link needs strengthening, as 'real' nature becomes associated with areas far away from our urban living environments. Cronon (1996b) mentions the risk of not taking responsibility for the environments we live in, as we prefer 'wilder' nature and the urban-industrial civilisation is not really our home. Idealising a distant wilderness often means not idealising the environment in which we actually live, the landscape that for better or for worse we call home. Urban green spaces and city forests in particular thus have an important role to play in (re-)connecting society and nature also in the places we work and life. Cronon (1996a, b) stresses that the 'otherness' of nature, the nonhuman can also be experienced close to home. This is not a second-rank experience. It is necessary to take the positive values associated with wilderness and bring them closer to home. In order to make a home in nature, the focus should be more on wildness than on wilderness.

City forests often embody more than other green spaces the wild side of town, as outlined above. But when developing their important role as (wilder) nature nearby, it is important to take the topics discussed before to heart. Central questions are: whose nature, whose 'wild side' are we talking about? How do these people define 'nature', 'wild', and 'forest'? What are their preferences? How do we balance the many different views of nature, for example also with maintaining or developing the forest's plant and animal life? Kilvington and Wilkinson (1999) stress that in activities such as biodiversity restoration in urban environments, due consideration must be given to people's values about and attitudes to urban vegetation within the landscape.

When developing the future 'wild' city forests, studies of people's preferences provide a useful base. Several interesting studies have been discussed above. Woodland visitors in Redditch, United Kingdom expressed a clear preference for mixed woodlands over coniferous woodlands. Also, a certain minimum size (2 ha) was considered as needed for offering a 'real' woodland experience (Coles & Bussey, 2000). Visitors of the recreation area Wienerberg in Vienna, an artificial recreational landscape established on a former industrial site, were highly appreciative of the 'naturalistic approach' followed when the area was developed during 1983–1995 (Arnberger & Eder, 2007). In general, more natural settings are preferred over urban settings (e.g. Herzog, 1989; Relf, 1992). Between urban settings, those with most nature are preferred. Unmanaged nature is preferred less than landscaped areas. Finally, in all cases, trees are highly valued components. A study by Kaplan (2007) of employee's reactions to nearby nature at their workplace also showed that natural settings are preferred – especially if they are of a walkable scale – over working environments with major buildings or parking areas. The presence of large trees was one of the factors leading to higher appreciation.

Although nature and 'wild' are preferred, research also indicates that it is important to have some 'cues' to management and care. Nassauer (1995, cf. Kaplan, 2007) mentions that more natural landscapes mostly have a rougher, wilder appearance and look 'messy', and therefore they need 'cues for care' to enhance their acceptability. Respondents in the previously mentioned study in Christchurch, New Zealand, accepted 'wilder' vegetation in their city, although they did feel that many areas looked 'too messy' (Kilvington & Wilkinson, 1999). A study in the Canadian city of Niagara Falls showed that visual preferences of decommissioned industrial lands, which are often seen as 'messy', can be significantly improved when 'vernacular cues to care' are added, such as bird boxes (Hands & Brown, 2002).

The terms 'wildspace' and 'wildscape' are increasingly being used in a context of urban green space (e.g. Simson, 2007b). Work by the OPENspace Research Centre of Edinburgh (Thompson et al., 2006) indicates that young people seem to have a particular need for 'wild adventure space'. This type of space is defined as outdoor space where young people have some form or freedom in terms of activity and experience. It has been found to offer multiple benefits, as described in Chapter 10 'The Forest of Learning'. Young people themselves state that wilder adventure space offers them a breathing space away from family or peers, a place that offers risk and challenge, to have a good time with friends and to really relax and feel free. The Wild Side of Town thus connects to The Great Escape.

An interesting case of promoting the 'wild side of town' is that of the 1,100 ha Sihlwald of Zurich (Irniger, 1991). During the 1980s, the plan was conceived to develop this forest area into a 'wilderness resort' (e.g. Seeland et al., 2002). Timber cutting was stopped (the forest had long been a 'fuelwood forest' for Zurich) and natural values were favoured. Presently the larger part of the Sihlwald is no longer managed. The forest should give the people of Zurich the opportunity to experience a 'natural forest', a wilderness area and to observe natural processes. The ambitions are illustrated by the name of a recent exhibition at the Sihlland nature centre: 'Vom Urwald zur Nutzwald und zurück' (From primeval forest to production forest and back). The aim is for the Sihlwald to obtain the status of a national 'Naturerlebnispark' (Nature Experience Park). According to new legislation for protected areas in Switzerland, these types of parks are 'small brothers' of national parks. They are situated near urban centres, are relatively small scale (but at least 6 km^2) and offer recreational and educational opportunities based in natural areas (Bachmann, 2006; Zürich will Naturpark, 2006). A study showed that, in spite of some initial resistance, most major user groups support the idea of establishing a 'peri-urban wilderness' in spite of restrictions that are enforced on recreational activities (Seeland et al., 2002).

The developments in the Sihlwald are an example for similar efforts to develop 'urban wilderness' elsewhere. In Berlin, for example, the environmental NGO BUND strives for wilder forests and nature under the slogan 'Wildniss in Berlin'. The growth in adventure holidays, ecotourism, visits to national parks and the like shows that there is a growing demand for wilderness experiences. Wild forests in the style of the Sihlwald can help urbanites keep in touch with nature. Moreover, there is a cultural need for wilderness, also in our cities, in a time when people (at least seem to) dominate nature (BUND Berlin, 2007).

Another interesting project to develop 'wilder' nature in cities is the British Wildspace! project (Preston, 2007). Supported by a lottery grant during 2001–2006, this scheme was dedicated to the creation and improvement of so-called Local Nature Reserves (LNRs). English Nature, an NGO, has the aim for more people to enjoy wildlife as part of their daily lives. LNRs are thus developed on the doorsteps of many towns. Local people have taken an active interest in creation and development of LNRs. During the time of the grant, 330 new LNRs were created, 77,000 volunteers were involved, as were more than 200,000 schoolchildren.

A wilderness experience can also be offered on abandoned brownfield sites. Lafortezza et al. (in press) show that 're-mediated' brownfield sites offer good opportunities from an ecological as well as an aesthetic perspective. An example of this are the new 'wild industrial forests' of the Ruhr area. These new 'wilderness areas' in the cities offer a contrast to dense urbanisation and a solution to the difficult financial situation of the local municipalities that cannot afford to create classic green spaces like parks. Instead of building a park, funds have been invested in foresters that look after the wilderness areas and people visiting them (Kowarik, 2005; Lohrberg, 2007).

Hiss (1990) states that, as people seek wilderness experience, it is important to provide mini-adventures in and near major cities (Fig. 8.3). This is true in the case of the industrial woodlands of the Ruhr area, where opportunities are offered for a wide range of activities. When people can stay closer to home, they will do so. This can then take some pressure off national parks and other more remote, often intensively used, nature reserves. Near-urban areas such as the Solleveld dune area in The Netherlands, for example, attract so much recreational interest from neighbouring cities such as The Hague that heavy regulations are imposed. A permit system restricts the number of visitors to an annual maximum. In this way, the areas' natural values and water protection functions also can be safeguarded (Duinwaterbedrijf Zuid-Holland, 2007). That recreational use of city forests and other 'wilder' nature areas in and near cities will need to be regulated is also described by Sapochkin et al. (2004) in the case of National Park Losiny Ostrov in Moscow. Studies show that massive recreational use of the Park have had quite a negative impact on the area's natural values and that the recreational carrying capacity is often far exceeded.

These examples of commendable efforts to make city forests 'wilder'. But barriers will have to be overcome, as already mentioned before, especially when it comes to people's preferences and reservations. Worpole (2006) writes that in the case of Greater London, many residents are ambivalent about nearby uncultivated wildland. Some residents are fearful of borders to semi-rural hinterland and put up boundary markers in some form.

Fig. 8.3 Wilder city forests can provide adventure at people's doorstep, as in this example of a nature area in the Swedish city of Joenkoeping (Photos by the author)

Innovative city forest management that promotes natural values and involves local communities holds the key to the future wild side of town. As Korthals (1998) states, all nature today is linked to people and management, as domestication of nature has progressed over time. Naturalness and management need one another. In urbanised societies, nature can no longer be left alone, for example when looking at the need to manage risks. Nature needs to be helped on its way, in the words of Fairbrother (1972). Or, according to Simson (2007b), who describes the case of the Südgelände Nature Park in Berlin: 'design' of wilderness is both possible and desirable. This is an interesting lesson for developing city forests as Wild Side of Town.

Chapter 9
The Healthy Forest

The World Health Organization (1946) defines health as the state of complete physical, mental and social well-being, and not merely as the absence of disease or infirmity. Thus health is seen in a rather broad way, explicitly including people's well-being. In spite of the advances made in medical and public health approaches, various forms of poor health have emerged in modern society. The majority of all causes of ill health, disease and premature deaths in the European Union and other parts of the industrialised world, relates to lifestyles, habits and environment (Nilsson et al., 2007). Contributing factors to poor health include an increasingly sedentary population, growing levels of psychological stress associated with urban living and contemporary work practices, and exposure to air pollution and other environmental hazards. Lack of physical activity and stress, for example, have led to increased occurrence of certain diseases where medication is perhaps only reducing the symptoms rather than combating the true cases of illness and reduced quality of life. Danish research shows that the proportion of the adult population that is severely overweight has increased significantly (e.g. Heitmann, 2000). Also young people are affected. Obese adolescents in Belgium tend to have a less positive attitude towards sports and participate less in sporting activities, thus worsening their health situation (e.g. Deforche et al., 2006). Obesity is not evenly distributed amongst all parts of society. Research in Austria showed, for example, that obesity is significantly higher amongst young immigrant girls as compared with other girls (Kirchengast & Schober, 2006).

Dealing with ill health is expensive, especially in comparison with prevention, as various studies have shown. Allender and Rayner (2007), for example, estimate that obesity-related ill health in the United Kingdom costs the country's National Health Service 3.2 billion GBP (about 4.4 billion EUR) on an annual basis.

It is clear that action needs to be taken. More focus is called for on alternative ways to prevent disease and promote human health and well-being. Moreover, people with disabilities and chronic illness increasingly demand a transition from institutional care to care in society (Nilsson et al., 2007).

Natural outdoor areas and natural elements such as forests, gardens and trees are known to provide opportunities for activities that enhance public health and well-being (European Landscape Convention, 2000; Stigsdotter & Grahn, 2002; Gallis, 2005; Nielsen & Nilsson, 2007; Nilsson et al., 2007; Worpole, 2007). Within

C.C. Konijnendijk, *The Forest and the City: The Cultural Landscape of Urban Woodland*

the larger resource of nature and green space, forests in and near cities have played an important role from a human health perspective. In the beginning, forests in general provided essential contributions to people's very subsistence. Later, city forests in particular started to act as a 'getaway' from often unhealthy city environments, offering opportunities for a wide range of (physical) activities. Hospitals and health centres have been situated in the peace, quietness and cleaner air of city forests as well. With present increases of stress, obesity and other welfare diseases, forest and nature are being 'rediscovered' once again as contributors to human health and well-being. City forests are central to this effort, as natural outdoor areas situated closest to where most people live and work.

This chapter starts with a more general, historical perspective on the links between (city) forests and human health and well-being. After this, the impacts of city forests on mental, respectively physical health will be discussed. Finally, ways to further promote the contributions of city forests to better human health and well-being will be considered.

Views on Nature and Health

The links between nature and health do not constitute a new phenomenon. According to Muir (2005), 'therapeutic landscapes', defined as landscapes that promote people's physical and mental health, have been created since biblical times. The author gives examples such as gardens with medicinal plants near abbeys and hunting parks of medieval castles, while the ancient Greek and Roman civilisations also recognised the potentially therapeutic workings of nature and landscape. Some landscapes that act as settings for human life have various therapeutic influences and have been preferred and 'reproduced' by people over time. The concept of 'therapeutic landscape' was developed further by Gesler during the early 1990s – see Gesler (2005) and Milligan and Bingley (2007) for an overview. Gesler also stressed that certain natural and built environments can promote mental and physical well-being. In these landscapes, environmental, societal and individual factors work together to promote health. The importance of everyday landscapes, such as those urban landscapes close to people's homes, should not be overlooked in this respect.

An important factor in the recognition of the positive effects of open spaces on public health was the ongoing industrialisation of society. Tam (1980) describes how in Britain, industry gravitated to logical, convenient places where a large working population was based or could be assembled. These industrial towns expanded and a large working class emerged, offering a stark contrast to the heavily landscapes villas and mansions of the newly affluent. The 1848 Public Health Act recognised the dangers to human health caused by the industrial city and the speed with which it grew. Large parts of the population were living in very bad and unhealthy conditions. However, initially the interventions of municipal authorities aimed at ameliorating this situation were hesitant and partial at best. In fact, philanthropists, preservation societies and industrialists often took the first measures, such as establishing new public green space (see also Chapter 5 'The Forest of Power' and Chapter 6 'The Great Escape'). Industrialists had much to gain from workers being healthy and satisfied. Together with other influential individuals – such as city mayors (Hennebo & Schmidt, s.a.) – they championed green-space causes. So-called metropolitan garden associations facilitated the process in Britain and influenced government intervention, which grew with the revised Public Health Act of 1875. This increased involvement led to establishing public parks which were being regarded as key elements to improve the 'debilitating environment' of industrial cities (Tam, 1980).

Worpole (2007) mentions how starting from the 19th century, the British government recognised the role which public parks could play in the 'physical and spiritual renewal' of the urban classes. The Victorian government saw the provision of parks as symbolic of a wider commitment to the public good. Physical well-being was clearly in focus, but so was an agenda of character formation and citizenship (see also Chapter 5 'The Forest of Power'). From the 1890s onwards, leisure activities such as walking, cycling, camping and trips to the countryside and seaside were associated with political and health reform. In Britain, this reform – which was promoted by grass root organisations – was characterised as 'the art of right living'. In Germany, where movements such as those of the Wandervogel pursued similar reforms through promoting outdoor hiking, the movement was called 'Lebensreform' (life reform).

Elsewhere in Europe, links between health, outdoor recreation and green space were also recognised during the 19th century. Urban green space could be used as 'lungs' of the city (Lawrence, 1993) and as breathing spaces (Clark & Jauhiainen, 2006). Like in Britain, those in favour of public parks and gardens also saw these as providing the additional benefit of reducing social conflict in the city (Lawrence, 1993), as they helped to discipline the urban masses into a 'world of respectable cultural values' (Clark & Jauhiainen, 2006). In France, the healthy aspect of walking for body and soul was stressed when the green structure of Paris was developed (e.g. Kalaora, 1981). Also during the 19th century, Dutch academics stressed that parks could make people happier and healthier (Dings & Munk, 2006). Forested recreation areas were established near large Dutch cities for recreational and 'sanitary' purposes (e.g. Buis, 1985; Van Rooijen, 1990). The German Volkspark principle also stressed the importance of extensive green areas that allowed for a range of active forms of public recreation.

Reeder (2006b) describes the developments in London during those days. The value of open spaces for people's health and quality of life gradually became focused on all Londoners rather than only on the elite. Various leading figures in society stressed that public green space was good for 'proper' amusement and health of the working class. Reeder (2006b, p. 43) writes: "Nature, like art, was thought to have a morally beneficial influence as well as recuperative powers." City forests and commons also featured in this development, although planners and designers focused on bringing more 'artifice' to nature with specimen trees, flower beds and the like.

During the early 20th century, stronger city administrations, sometimes helped by national governmental policy, actively engaged in making their cities green and healthier. Stockholm's park policy, for example, had great success during those days, and particularly during the 1930s. The city's public green space was expanded to over 4,000 ha (Nilsson, 2006a). Parks were clearly seen as important means to improve public health. Similar to elsewhere in Europe, Stockholm's green policies included a Garden City movement that considered it important for people to live among green space. This was also the time of influential city gardeners, such as Holger Blom in Stockholm (see also Treib, 2002). Inspiration was taken from other parts of the world, including the American City Beautiful Movement.

According to Worpole (2007), the links between open space, outdoor recreation and public health were given less government attention – at least in Britain – during the second part of the 20th century. Health was increasingly regarded as a private matter. However, during recent years, nature's health effects have come to the fore once again, not in the least because of the changes in society sketched in the introduction to this chapter.

How about the role of forests in these developments? Interestingly, forests had not always been seen as healthy places. During the 18th and even 19th century, they were assumed to harbour unhealthy air, being damp and unclear. Many forested countries were thus considered to have 'bad air' (Lehmann, 1999). This may explain the focus on public parks rather than 'wilder' nature in initial efforts to link green space and public health. However, during the 1960s, health was specifically linked to forest in a more positive way, for example at a congress of the International Union of Forest Research Organizations (IUFRO) in 1967. Participants discussed the question how necessary outdoor recreation in forests was – within the total scope of leisure time – as an 'adjunct' to urban life. Times were regarded as becoming more stressful, for example due to higher-than-ever family expectations. Outdoor recreation was debated as a way to improve mental and physical health and reduce crime and delinquency (Arnold, 1967).

Promotion and development of the linkages between outdoor environments (including forests) and public health and well-being have been hampered by the lack of scientific evidence on health effects and mechanisms at work. Although the general feeling is that nature is good for people's health, nature is not yet widely used for health promotion by public authorities, partly because of the lack of 'hard facts'. Green space nearby is still often seen as a luxury rather than a necessity (or a public service), especially in urban areas where the competition for land is fierce. Many aspects and mechanisms in nature-health linkages are still largely unknown, although we intuitively may feel that nature promotes human health and makes us feel better and happier. But what are the causal relationships – if any – between views and use of natural landscapes and our health and well-being? Which types of landscapes have the most positive effects? Slowly but steadily, empirical knowledge is being accumulated in a wide range of scientific fields, including environmental psychology, landscape architecture, forestry, and epidemiology (Bonnes et al., 2004; Gallis, 2005; Nilsson et al., 2007; Velarde et al., 2007). Recent studies have looked, for example, into accessibility and use of nature on (self-reported) human health and well-being. Studies that compared health indicators with access to green spaces in Denmark, England and The Netherlands found that both health and well-being were better among people who regularly visited nearby nature and green spaces (Maas et al., 2006; Mitchell & Popham, 2007; Nielsen & Hansen, 2007).

City Forests and Physical Health

Since ancient times, forests have been a source of essential foodstuffs, thus contributing to people's very survival (see Chapter 4 'The Fruitful Forest'). Trees and other vegetation have also been used in traditional, modern and alternative medicine as sources of pharmaceuticals and other chemicals (e.g. Perlis, 2006). In some cases, city forests in especially tropical countries are still a source of alternative medicine (Fig. 9.1). A recent study of Ayer Hitam Forest Reserve near Kuala Lumpur, Malaysia, for example, indicated that medicinal plants are still being extracted by indigenous people living near this urbanised forest (Konijnendijk et al., 2007).

A recent European report on the positive linkages between outdoor environments and human health and well-being (Nilsson et al., 2007) considered, among other issues, the role of forest products in current medicine. As part of complex chemical defence systems, trees include bio-active, protective substances such as flavonoids, lignins, stilbenes, terpenoids, phytosterols, fatty acids and vitamins that have (potential) health-promoting effects. Bio-active tree compounds can be used as nutraceuticals, that is, a combination of nutrition and pharmaceuticals. These can contribute to public health as ingredients in dietary supplements and health-promoting ('functional') foods, and as pharmaceuticals. Xylitol products, for example, promote dental health (e.g. Mickenautisch et al., 2007), while substances such as taxol (derived from yew, *Taxus baccata* L.) have been found to be powerful antioxidants that can be used in the fight against different forms of cancer (e.g. Walsh & Goodman, 2002).

By acting as a biological buffer, trees and forests also affect human health and well-being by helping to moderate the effects of other physical environmental factors. Trees and forests can filter potentially harmful air pollution and solar radiation, they provide natural shelter against the wind and they help to cool and moisten the air (e.g. Nowak, 2006). Hiss (1990, p. 118) cites Parisians who state that the city's famous golden, glowing light would not exist without the Bois de Boulogne and Bois de Vincennes blocking out dust on both sides of the city centre. These impacts of city forests on climate and environment could become even more important due to climate change, as this ongoing process also impacts human health. Comrie (2007) describes how health problems could be worsened by climate change, as this is likely to lead to increase of precipitation, temperatures and their variability. These changes are expected to result in an increasing number of diseases, such as heat-wave related problems. The European summer of 2003, with its many heat-related deaths, serves as an example. Other results can be increases of air pollution related diseases, aeroallergen-related diseases, fungi- and mold-related diseases, water- and food-born diseases and influenza. It is clear that nature and

Fig. 9.1 In some countries, local inhabitants still use city forests for collecting medicinal plants. This Arab man shows his harvest from the Rosh Ha'ayin city forest in Israel

trees are linked to many of these developments. For example, a higher tree cover in cities could enhance shading and reduce temperatures, thus buffering temperature increases (e.g. Graves et al., 2001; Nikolopoulou et al., 2001).

But not only through their products and environmental services provided do city forests and trees promote human physical health. As forests and other natural environments are seen as more attractive than built environments (e.g. Velarde et al., 2007; see also Chapter 8 'The Wild Side of Town'), green areas may stimulate residents to undertake healthy physical activities such as walking or cycling. Residents may also choose these activities as a mode of transport and spend more time in city forests and other green spaces (Groenewegen et al., 2006). This has been recognised for a while. A Swedish exercise physiologist suggested establishment of network of trails and installations for physical health in Norra Djurgården in 1957 (Schantz, 2006).

That regular physical activity in nature improves people's physical health is also shown in a literature review by Jorgensen and Anthopoulou (2007). Walking is known, for example, to counteract or prevent obesity and coronary heart diseases. Living in areas with access to walkable green spaces can positively influence longevity. In the United Kingdom, physical inactivity was found to be directly responsible for 3% of morbidity and mortality, resulting in an estimated cost to the National Health Service over 1.5 billion EUR annually (Allender et al., 2007). Physical activity has been associated with reduced emergency admissions and hospital stays amongst socio-economically disadvantaged older adults, as well as with reductions in health care charges of amongst elderly retirees. Nielsen and Hansen (2007) also present ample evidence that physical inactivity results in high risk of premature development of chronic diseases.

In another recent study, Schriver (2006) looked at the use of woodland in rehabilitation processes and physiotherapy after physical injury. Patient logbooks, observations and patient interviews indicated that the use of woodland had a positive effect, for example in terms of understanding patients' changed situation after injury and by influencing their senses, moods and feelings as important factors in the personal rehabilitation process.

Urban green space, including city woods, can also help promote better physical health at the workplace. Business parks with primarily offices and knowledge-based industry seem most promising here, as offices are easiest to combine with recreational use and personnel are mostly involved in sedentary, mental types of work (Jókövi et al., 2002). In The Netherlands, for example, 73% of all employees consider themselves as having sedentary jobs (Hendriksen et al., 2003). Encouraging employees to walk during their lunch breaks could help improve their physical as well as mental health, as research has shown, although so far only few studies have looked into this particular topic. With this in mind, the Dutch government recently carried out a national survey of workers. This showed that 34% of the 73% employees with physically inactive jobs regularly used their lunch breaks for walking. Attractive local environments were considered of crucial importance for getting people out. This is confirmed by another Dutch study which focused specifically on employees working at 'green' business parks, i.e. workers who had such an attractive environment at their availability. This study showed that 89% of the employees said that they used the surrounding green environment occasionally and 92% appreciated the green and recreational environment. The typical period of use was 16–30 minutes, while getting fresh air, relaxing and being in nature were the main motivations (Jókövi et al., 2002).

Links between hospitals and other health care institutions and green areas have existed for a while (Fig. 9.2). During the 19th century, for example, several hospitals were established in the Bois de Vincennes of Paris, as it was believed that the forest environment and clear air would be beneficial for patients (Derex, 1997b). A much-cited study by Ulrich (1984) showed that patients recovering from gal bladder operations were able to leave the hospital quicker when they had a 'green' view

Fig. 9.2 Many hospitals and other healthcare institutions are located in or near city forests, as here in the case of Konstanz, Germany (Photo by the author)

from their hospital window. Hospitals and institutions have often used green environments and 'healing gardens' for health promoting purposes (e.g. Stigsdotter & Grahn, 2002). In some cases, health care institutions have teamed up with forest and nature managers, with the aim to promote patients physical and mental health by involving them in management activities. A Dutch study (Oosterbaan et al., 2005) indicated a growth in the number of former addicts, psychiatric patients and handicapped being involved in the management of forest, nature and landscape, with between 1,500 and 2,000 clients of health organisations working for nature organisations and the national forest service.

Children are a special group when it comes to links between nature, forests and physical health. American author Richard Louv (Louv, 2006; Talking with the author, 2007) stresses the importance of child's play outdoors, as direct contact with nature can have an important role in healthy child development and may be part of the solution to such health issues as ADHD, child obesity, stress and decreased creativity and cognitive functioning (cf. Taylor & Kuo, 2006).

The elderly comprise another important group. A study amongst British elderly found that those who live in a 'supportive' environment tended to walk more, and high-level walkers were more likely to be in good health (Sugiyama & Thompson,

2007). Neighbourhood environments should thus be designed in such a way that they provide opportunities to be active. Moreover, more places need to be offered where people can meet with each other and enjoy nature.

Forests and nature can also be used as a symbol for (fighting ill) health. In Flanders, Belgium, the annual 'Boompjesweekend' (Small Trees Weekend, a national tree planting event) is organised in collaboration with the national campaign to fight cancer. People can buy a 'tree package' and help fund the campaign. In 2007, tree planting was undertaken in six new forests, all situated close to cities (Cnudde, 2007).

Finally, linkages between nature, forests and physical health are not always positive. In Chapter 3 'The Forest of Fear', several negative aspects of city forest use are briefly mentioned, such as animal-borne diseases and fear of wild animals. Elderly people in the United Kingdom have expressed fears of getting hurt in forests, for example when stumbling over a root or other unevenness on paths (Jorgensen and Anthopoulou 2007). Thompson and Thompson (2003) focus on allergies, stating that the number of potentially allergenic plants has grown rapidly in urban areas, as the diversity of plants increases. Some of the most commonly planted trees in urban areas are known to be the greatest producers of pollen.

City Forests and Mental Health and Well-Being

Modern times have brought about an increase in stress and mental disorders. In the United Kingdom, one in ten children between the ages of 5 and 16 had a clinically diagnosed mental disorder according to 2004 statistics (Milligan & Bingley, 2007). Nielsen and Hansen (2007) cite data from the National Institute of Public Health in Denmark, which indicate that 44% of the Danish population experienced stress in the year 2000, compared to 35% in 1987.

Contact with nature can have a powerful therapeutic or preventative effect on many people. Exposure to green space consists of direct physical exposure and the psychological processes through which exposure influences health and well-being. Apart from improving physical ability, as discussed above, contact with nature can reduce stress and help improve mental ability as well. Descriptive epidemiological research has shown a positive relationship between the amount of green space in the living environment, not only with physical health and longevity, but also with mental health. Looking out on, and being in, the green elements of the landscape around us seem to affect health, well-being and feelings of social safety (Groenewegen et al., 2006).

Velarde et al. (2007) reviewed studies linking landscapes and health effects, focusing on the positive health effects of viewing landscapes. The authors found that, in general, natural landscapes have a stronger positive health effect than urban landscapes. Most studies confirm this, but not much information is available about the magnitude of the differences. There are three main health effects of viewing landscapes: (1) short-term recovery from stress or mental fatigue, (2) faster physical recovery from illness (see the previous section) and (3) long-term overall improvement of people's health and well-being.

Experimental research in environmental psychology has shown that a natural environment has a positive effect on well-being through restoration of stress and attentional fatigue. Two of the main theories behind this are Roger Ulrich's Stress Recovery Theory (e.g. Ulrich, 1984) and Rachel and Stephen Kaplan's Attention Restoration Theory (e.g. Kaplan & Kaplan, 1989; Hartig, 2004). The first theory states that natural scenes tend to reduce stress, whilst settings in built environments tend to hinder recovery from stress. Evidence suggests that exposure to natural environments may reduce feelings of anger, frustration and aggression. In turn, this may enhance feelings of social safety, and even reduce actual rates of aggressive behaviour and criminal activity (Groenewegen et al., 2006). The second theory focuses on the restorative effects of natural environments on human mental fatigue. Restoration is the process of renewing physical, psychological and social capabilities diminished in ongoing efforts to meet adaptive demands. Stephan Kaplan has stated (in Gallagher, 1993, p. 214) that nature experiences can play a vital role in helping people manage the problem of inner-city settings, which are often environments of over-stimulation. Kaplan mentions the case of North-America's urban poor, who are under 'all-out assault' from their hectic urban environments, meaning that it is difficult to focus one's attention. People like this are in great need of restorative experiences, such as those offered by urban nature.

Natural landscapes have been used to treat burnout syndrome and depression. As discussed earlier, therapeutic landscapes relate to the use of particular places for maintenance of health and well-being. Traditionally these were regarded as healing places (e.g. Gesler, 2005, also Velarde et al., 2007), but are now defined broader. The term 'restorative landscapes' has also been used, referring to therapeutic landscapes that help restore emotional and physical well-being (Milligan & Bingley, 2007). Psychiatric patients, for example, have often been treated in a quiet (and often rather isolated) natural setting. Wytzes (2006) describes a trend in The Netherlands to move psychiatric patients from the traditional remote, quiet treatment areas in nature to busier city environments, in order to integrate the patients into 'normal life' and make them more independent. The traditional, natural healing environment became seen as a type of 'prison' during the 1970s. However, the move of patients to the city often led to a decline in health and well-being. Now, ways are being explored to combine a natural treatment setting with a certain level of patient autonomy.

The earlier-mentioned report by Nilsson et al. (2007), which was commissioned by the European Cooperation in the field of Scientific and Technical Research (COST), states that so far, experimental research on stress reduction and attention restoration has mostly focused on short-term stress reduction, with stress being induced within the experimental setting. Furthermore, usually a crude distinction is made between a natural and a built-up environment, often represented by slides or videos. However, based on this line of research it is difficult to say: (a) what the (size of) long-term health benefits of exposure to nearby nature in the residential or working environment will be, (b) what type of nature will work best, (c) how much of this type of nature is needed, and (d) whether there are additional requirements that should be fulfilled.

Being situated closest to where most people live, urban green space such as city forest plays an important role within natural landscapes for the promotion of people's health and well-being. Moreover, city forests feature among people's favourite places (e.g. Lipsanen, 2006). Korpela and Hartig (1996) found that favourite places offer experiences that are associated with restorative environments. Visitors of the Vondelpark, the most visited urban park of Amsterdam, see experience of nature in the urban environment as a source of positive feelings and beneficial services. Green areas such as the Vondelpark are perceived to have many social and psychological benefits; users see the park as a restorative environment (Chiesura, 2004). In Helsinki, green areas including city forests have long been seen as having a positive influence on the 'mental stability' of urban inhabitants (Lento, 2006). In the United Kingdom, many people have had restorative experiences in urban woodlands, ranging from stress-relief to opportunities to deeper reflection and self-regulation (Bussey, 1996, cited by Jorgensen et al., 2007). Also in the UK, a recent study carried out for mental health charity 'Mind' compared a walk in a country park with one in a shopping mall (Mind, 2007). Seventy-one percent of the respondents reported a decrease in levels of depression after a walk in the park, whilst 22% reported an actual increase in levels of depression after the shopping walk. Additionally, regular participants in Mind's network of local associations reported that a combination of nature and exercise had the greatest positive impact on them. Ninety-four percent of the respondents said that green activities had benefited their mental health, lifting depression.

Young British adults between 16 and 21 years old were the focus of a study by Milligan and Bingley (2007). The researchers found that woodland often acts as a therapeutic place for young people. It provides a place where to go when feeling upset and to sit and recollect one's thought. The therapeutic nature of woodland was attributed, in part, to the size and age of trees, but also to their perceived qualities of safety, protection and calmness. Trees were seen as symbols of stability and continuity (see Chapter 2 'The Spiritual Forest'). The study also pointed at the long-term effects of childhood play on subsequent mental well-being in young adulthood. Young adults who had frequently played in woodland when they were children were more likely to see these as therapeutic landscapes.

A survey in Denmark assigned by a washing detergent company surveyed 630 parents with children between 5 and 15 years old, with the aim to identify the families' relation to dirt and physical activity. The study indicated that 83% of the parents think that their children become less stressed when spending time outdoors. However, children were found to spend more time on indoor than outdoor activities. Most parents think that computer games and Internet chat are the reason for children staying indoors. More than two-thirds of the Danish kids have a television and/or computer in their room (WEBPOL, 2006).

A special case of mental health impacts of nature is offered by wilderness experience. Nature and wilderness are different sizes, for example in terms of psychological and restoration effects (e.g. Kaplan & Kaplan, 1989; Gallagher, 1993).

Wilderness experience, as also described in Chapter 8 'The Wild Side of Town', has been associated with a range of spiritual and transcendent experiences that provide benefits such as greater self-confidence, sense of belonging to something greater than oneself and renewed clarity on 'what really matters' (Kaplan & Talbot, 1983; Relf, 1992; Gallagher, 1993). Knecht (2004) mentions that wilderness experience can also take place in areas close to cities perceived as wilderness or wilder nature. City forests are an obvious place to offer this type of experience. Wilderness experience may then provide stress-reducing and attention-restoring benefits of everyday nature in a more long-lasting way. Interestingly, although wilderness is often connected with individual experience of the 'grandness' of wild nature, it turns out that many of us experience wilderness together with others. In North America, for example, only 3% of visitors to large nature parks come alone (Gallagher, 1993).

Although linkages between use and views of the natural landscape seem largely positive, past discourses such as those of the forest decline ('Waldsterben') debate in Germany show that negative feelings can also be evoked when people look at forests or other natural landscapes (Lehmann, 1999). Feelings of death and loss can be associated with, for example, forests being perceived as 'dying'. Work by Perlman (1994) confirms this. He interviewed inhabitants of Florida, USA about their relationships to trees after a severe hurricane had hit, toppling over many city trees. Also, threats to local forests and trees have resulted in fierce protests and emotions running high in many cases across Europe. Chapter 12 'A Forest of Conflict' gives examples of conflicts and emotional protests involving city forests.

Promoting the Healthy Forest

The past sections have shown that natural landscapes, including city forests can have positive impacts on people's mental and physical health, as well as general well-being. Moreover, natural environments are preferred over urban ones. Within urban areas, more natural settings are clearly preferred. What is the role of nearby nature and especially of the city forest within these impacts? Some of the answers to this question have already been given, but it is useful to explore this issue a bit further. According to Hartig (2003, cited by Milligan & Bingley, 2007), in a time of burgeoning health care costs and declining environmental quality, public health and health promotion strategies that have a natural environment component to them may have particular value for an increasingly urbanised population. People make most use of outdoor spaces that are nearby, engender a sense of security and ownership, and are attractive (Nilsson et al., 2007). Nearness to green areas and (as a result) higher numbers of visits to green space are associated with lower levels of (self reported) stress. Moreover, people with a garden of their own were found to be less affected by stress (e.g. De Vries et al., 2003; Nielsen & Hansen, 2007). Nearness does not even have to mean actual physical use of forests and green space, as the work of Ulrich and others have shown. A recent study in South Korea indicated that office workers with windows looking on forest scenery have lower levels of stress and higher job satisfaction (Shin, 2007).

Nearby nature such as city forests thus plays a very important role in the linkages between outdoor environments and human health and well-being. 'Nearby nature' consists of the natural elements and features that people encounter in and around those settings of everyday life in which they spend much of their time, including residential settings, the workplace and schools (Nilsson et al., 2007). Non-threatening nature is typically liked. We have also seen, however, that some form of 'wilderness experience' can have a positive effect on mental health and overall well-being. People commonly want to have access to nearby opportunities. A short nature experience can produce benefits that typically have a brief duration, whilst proximity to nature can yield benefits that stand as cumulative effects of repeated brief experiences (Fig. 9.3).

Fig. 9.3 Spending time in urban green spaces, such as city forests, can help reduce stress. People have been found to become more concentrated and focused after visits to green space (Photo by the author)

How can experiences like these be promoted? How can nearby nature such as city forests make even more contributions to the health and well-being of city dwellers? Of course specific actions to be taken depend on the precise relations between a certain type of green space and different types of health impacts. But as discussed earlier, ther is a lack of knowledge, particularly about the mechanisms at work. Obviously using natural elements to catch fine dust will lead to a different optimal green structure then using such elements to create a green oasis to relax and recover (Groenewegen et al., 2006).

Environments should be perceived as being away from people's daily surroundings, affording possibilities for fascination, being coherent in themselves and compatible with people's needs (e.g. Jorgensen & Anthopoulou, 2007). Woodland offers good opportunities for making people feel 'away' from the city, as shown in Chapter 6 'The Great Escape'. As mentioned in previous chapters, they may be relatively small to still offer some form of 'forest' experience (Coles & Bussey, 2000).

In order to deliver public benefit by encouraging activity in nearby natural settings, one needs to consider how to compete with people's other interests and calls on their time. Moreover, there is a need to motivate and sustain particular behaviours, design and promote appropriate places, identify and target different groups (i.e. segmentation) and fit these activities into the context of wider health objectives (Nilsson et al., 2007).

Accessibility is a key issue. Urban green space – including city forest – is not accessible to everyone, at least not to the same degree. For a significant part of urban society, access is limited because of physical barriers such as transport corridors, ownership issues, location, design and infrastructure (e.g. Nilsson et al., 2007). Informational barriers also exist where people do not know of areas, or of the possibility of using them. Access is considered in campaigns such as 'Copenhagen on the Move' ('En By I Bevægelse'). This major campaign was started by the Copenhagen city administration in response to the relatively low life expectancy of the city's inhabitants (Lee & Pape, 2006). Accessible and nearby green space plays a key role in the campaign, acting as a 'green pulse' of the city and offering a relatively inexpensive way to promote public health. An important target is that all residents of Copenhagen should live within 400 m of the nearest green space in line with a norm increasingly accepted across Europe (e.g. Van Herzele & Wiedemann, 2003). Efforts are made to also engage social weaker groups in various activities. The campaign encompasses a unique cooperation between 'green' and medical professions (Lee & Pape, 2006).

Green spaces must be attractive to users and have facilities appropriate to the main objective. In Denmark, a 'catalogue of ideas' was developed for public space, with suggestions for providing appropriate and inspiring settings and facilities to encourage physical activity. The catalogue looks at ways to promote activities such as play, outdoor sports, as well as outdoor education (Hansen, 2005). Differentiation is important this respect. It applies both to spaces and activities as well as to the social groups for whom they are provided. This also links to the topic of 'social health' which has not been given much attention here, but is dealt with in Chapter 11 'The Social Forest'. Being part of a group or a local community is important to people's health and well-being. Gathright et al. (2007), for example, identified positive social impacts of recreational tree climbing in Japan, linking positive experiences to family, friends and willingness to engage with local conservation initiatives. City forests as 'territories' and place also have their role to play in the promotion of social health.

Social health and the forging of stronger communities are part of the multiple roles that city forests and other parts of the natural environment play. Many of these roles, including the provision of environmental services, can have a positive impact on human health, as discussed earlier in this chapter. Jones and Cloke (2002) speak of 'therapeutic trees' that not only help reduce stress, but also clean the air and water resources, reduce solar radiation, reduce the risk of storm-water flooding, help reclaim contaminated land, and the like. City forests can thus be Healthy Forests in many different forms, contributing to all aspects of human health and well-being.

Chapter 10
The Forest of Learning

Over time, city forests have had various educational roles. Thus these forests can be considered as Forests of Learning. This chapter looks at 'learning' in a broad sense, including professional education, training and advancement, international exchange of ideas, as well as environmental education and keeping society in touch with nature and forestry.

City forests have been important testing grounds, for example for the introduction of exotic tree species as well as for the development of new approaches and methods in forestry. This role is described in the next section. The section after that considers how those owning or designing city forests have been inspired by woodlands, parks and developments abroad. City forests have also assisted forest managers in improving communication with the urban public. This is one of the roles of the demonstration forests that are also introduced in this chapter. The wider educational roles of city forests are also discussed. They offer a 'window to the world' and act as settings for environmental educational of particularly children and youths.

The educational roles of city forests, however, can be developed further, as is discussed in the final section of this chapter. Enhancement of the Forest of Learning can help meet the many challenges forestry and landscape management are facing in a rapidly changing and urbanising society.

City Forests as Testing Grounds

When rulers and the elite started to import exotic tree species, for example because of their ornamental properties, they often planted these trees in their estates. As seen in previous chapter, many of these estates would later develop into city forests with public access. The introduction and testing of exotic species was also part of the emergence and development of the science and profession of forestry. City forests played an important role in this process. They offered conveniently located – i.e. close to cities and educational institutions – testing areas, for example to determine the suitability of certainty tree species for timber production and the pros and cons of different silvicultural regimes.

In Europe, the introduction of exotic tree species started during the late 16th century and early 17th century (Spongberg, 1990). Private collectors, garden societies and botanical gardens were interested in collecting and testing new species. The Tradescant family of London, for example, collected plants from all over Europe as well as from the new colonies. They also helped introduce exotic European trees to North America. During the 18th century, an increasing number of Asian tree species, such as ginkgo (*Ginkgo biloba* L.) and golden rain tree (*Koelreuteria paniculata* Laxm.), were brought to Europe. The Dutch East India Company, then a dominating force in the trade between Asia and Europe, was especially influential in this introduction process. Blok (1994) describes how different exotic tree species reached what is now The Netherlands during the 17th and 18th century. This started with horse chestnut (*Aesculus hippocastanum* L.), followed by species such as London plane tree (*Platanus orientalis* L.), bald cypress (*Taxodium distichum* (L.)Rich) and cedar of Lebanon (*Cedrus libani* A.Rich). By the middle of the 18th century, many estates and botanical gardens in The Netherlands and other trading nations had very extensive tree and plant collections.

The introduction of exotic trees and plants in Europe reached its peak during the late 18th and early 19th century. In the Victorian era, the British aristocracy in particular was interested in introducing new, exotic species and supported the efforts and expeditions assigned by private organisations such as the Horticultural Society (e.g. Spongberg, 1990; Muir, 2005). This was the era of great botanical explorers such as Scotsmen Archibald Menzies and David Douglas. In their book 'The Tree Collector', Lindsay and House (2005) provide a fascinating account of the life of the latter. Douglas was sent to North America on several journeys by the Horticultural Society to find ornamental and other plants to be introduced to Britain. He was highly successful, giving his name to, for example, Douglas fir (*Pseudotsuga menziesii* (Mirb.)Franko).

Korthals (1998) provides an interesting perspective of three periods in Western society's attitude towards exotic tree and other species. The 18th century saw an appreciation of geographic origin. The 'acclimatisation movement' of the 19th century moved species all over the globe, in line with imperialistic thought. Efforts were facilitated by better transport conditions. Species were to be used there where they could bring benefit. The 20th century saw a counter movement against this, particularly within political movements such as national-socialism. Native species were defended to such extent that they were even 'bred back', as in the case of the Heck cattle (which resembles the extinct aurochs, *Bos primigenius* Bojanus).

Sometimes city forests themselves 'bred' new trees. A hybrid cypress, x *Cupressocyparis leylandii* 'Robinson Gold', was first identified at Belvoir Park Forest near Belfast. This notable tree became widely propagated and was chosen for the code of arms of local Castlereagh Borough Council (Simon, 2001, 2005).

Exotic trees were not only introduced to private estates. Gradually arboretums became of increasing importance when the scientific study of trees and vegetation boomed. Three types of arboretums were set up, often as part of wooded estates in or near cities. In so-called geographical arboretums, trees are brought together from one or more specific geographic regions. Arboretums that emphasise collections

Fig. 10.1 The Tervuren arboretum in the Zoniënwoud of Brussels is a geographically organised forestry arboretum. The illustration shows a map of the arboretum **(a)** and an impression of the area's scenery **(b)** (Photos by the author)

generally have the aim to bring together as many (special) species as possible. Forestry arboretums, finally, are set up to test the suitability of exotic trees to be used for forestry purposes (e.g. van der Ben, 2000). The latter became increasingly important over time, especially when forestry emerged as a profession.

Although the Tervuren arboretum in the Zoniënwoud of Brussels (Fig. 10.1) is a geographical arboretum, it has also been used as a forestry arboretum. Starting from the early 20th century, trees were planted according to their geographical origin and ecology, with emphasis on the Northern Hemisphere, so that trees could be tested for possible forestry use in Belgium (van der Ben, 2000; Geerts, 2002).

So-called 'rational forestry' and forest science which would influence forestry across Europe and the world, had their roots in late 18th-century Germany. The Berlin area was one of the centres where forestry came of age. The Berlin city forests proved to be particularly important, as they were used as experimental and educational areas (Cornelius, 1995). Several of the leading experts of German forestry were based at or worked in Berlin, including Georg Ludwig Hartig, the initiator of 'Prussian forestry'. Another key person, Friedrich Wilhelm Leopold Pfeil, was the first director of the 'Forstschule' (Forest School) of Berlin, which was established in 1821 and later moved to neighbouring Eberswalde, where it still based today. The forests near Berlin were used to develop new methods for forestry, such as for improving stands, regeneration and finding ways to optimise economic returns from forestry. The Berlin forest school tested exotic species such as black locust (*Robinia pseudoacacia* L.), (European) larch (*Larix decidua* Mill) and tree of heaven (*Ailanthus altissima* (Mill.)Swingle) in the Berlin forests, in order to explore their economic potential for German forestry.

German foresters were also invited to other countries to share and implement their knowledge. The Danish king, for example, employed German forester von Langen in 1762, with the assignment to test different tree species under Danish conditions, as well as to develop a suitable forest management system. A city forest, the former royal hunting ground of Jægersborg Dyrehave, became an important testing ground for von Langen. Present-day Dyrehave still includes an area called 'Von Langens Plantage' (Møller, 1990).

The experimental role of city forestry has gradually been expanded. This is not surprising, of course, as many forest faculties, schools and institutes are based in cities and have turned to nearby forest areas for their experiments and education. Kardell (1998) mentions the importance of Norra Djurgården in Stockholm for the development of the forestry profession in Sweden. Until 1977, this city forest hosted the first higher forestry education in Sweden, Skogshögskolan ('Forest High School'). The school used the surrounding city forests for testing, research and experiments. The Bahçeköy Forest Enterprise situated within the boundaries of Istanbul, Turkey, has been used for research and education purposes by the neighbouring Forest Faculty of the University of Istanbul. Part of the forest is a model coppice forest (Eker & Ok, 2005).

In some cases, universities, school and institutes even appropriated their own city forests for research and training. One example of this is the State Forest Technical Academy of St. Petersburg, which is situated in the middle of a large forest park which is also used for educational purposes (Anan'ich & Kobak, 2006). Yet another example of the close connection between city forests and universities is the 'Masaryk Forest' Training Forest Enterprise at Krtiny, which manages an area of over 10,000 ha and is linked to the Mendel University of Brno, Czech Republic. This forest area is located north-east of Brno, neighbouring the industrial town of Adamov. Education of students is one of the Masaryk Forest's main functions (Konijnendijk, 1999). In the case of Switzerland, Holscher (1973) mentions that the Swiss Federal Institute of Technology owned forests in Zurich which were used as testing grounds. Income from timber production helped to finance forestry research.

City forests helped to move the field of forestry forward, as testing and educational areas, but also as they were in the public eye most of the time. Many city forests were among the first to adopt principles of multifunctional forestry, as well as closer-to-nature forestry (e.g. Krott, 1998; Otto, 1998). This has been the case in Slovenia, for example. More than 1,500 ha of forests around the Slovenian capital of Ljubljana were proclaimed urban forest, resulting in a management focus on multiple use and closer-to-nature forestry (Pirnat, 2005). Dutch city forests have played an important role in the development of the country's forestry in, as well as in raising awareness about forestry issues (Paasman, 1997). During the second part of the 20th century, the deplorable condition of the Haarlemmerhout wood of Haarlem brought about the start of concerns about the general vitality of Dutch forests. A debate over management of a city wood called Baarnse Bos, initiated by an environmental group called 'Critical Forest Management' ('Kritisch Bosbeheer' in Dutch), led to more focus on 'close-to-nature' forest management. In Northern Ireland, Belvoir Park Forest has helped to raise public awareness about the production purposes of forestry (Simon, 2005). City forests in countries such as Belgium (Flanders), Germany, The Netherlands and the United Kingdom were also among the first to have their forest management certified by the Forest Stewardship Council and other organisations (Konijnendijk, 1999).

International Inspiration

City forests have been Forests of Learning in many ways. In Chapter 7 'A Work of Art', city forests are described as works of art by themselves, as they have been planned and (often artfully) designed according to the fashions of the time. International inspiration played an important role in this process. As early as during the Middle Ages, the European aristocracy was often in close contact and copied each other, for example in developing hunting domains and prestigious estates. They also did their utmost best to impress their noble guests. The Bavarian Kings, for example, took their visitors to their Forstenrieder game park near Munich. In a grand setting, hundreds of deer and wild boar were driven towards the guests, whilst they were comfortably seated on a grandstand, ready to shoot (Ammer et al., 1999).

Initially, as also discussion in Chapter 1 'Introduction', national cultures and characteristics dominated green space design. The Dutch had their tree-lined canals, the French their boulevards, and the English their squares (Lawrence, 2006). Gradually, however, fashions and styles were copied across Europe (e.g. Clark & Jauhiainen, 2006). This started during the Renaissance, when the example of the Italian aristocracy and merchants to develop villas with extensive parks (which included woods) was implemented in many countries. Later, the French baroque style was widely adopted for palace gardens and hunting forests. The same happened subsequently with the English landscape style during the period of Romanticism, when nature came to the fore. Large-scale, prestigious greening projects such as Hausmann's grand green structure works in Paris were also copied throughout Europe. The German Volkspark movement meant yet another wave of international 'fashion' in planning and design of city forests and parks. Allotment gardens also arose in many cities.

From the late 19th and early 20th century onwards, an international planning movement with strong links to North America emerged and extended its influence. Ideas such as those of Ebenezer Howard's Garden Cities and the New Town concept were implemented in Europe as well as elsewhere. City greening in Japanese cities such as its capital Edo (now Tokyo) became influenced by Western ideas during the 19th century (Treib, 2002). Before this, not much woods or other public green existed in the Japanese empirical city. Trees and green were mostly to be found within the walls of the empirical palace, while the only 'public' vegetation was to be seen over the walls of private aristocratic estates.

Another rather influential concept in Europe was that of the Stockholm Park System, also known as the Stockholm Style. As described in Chapter 8 'The Wild Side of Town', this concept was developed at the end of the 19th century under the leadership of city gardener Holger Blom. It was recognised internationally as a model for green space development (Treib, 2002). Its principles included that the park 'relieves' the city, that it is a place for outdoor recreation as well as a meeting place, and also that parks preserve nature and culture. The Swedish city park directors wished to preserve original topography and vegetation as much as possible. Many of the Stockholm green areas, for example, seemed like remnants of natural areas preserved within the city, although this was not always the case. Adapting to nature and to local natural conditions was the underlying principle, which was 'exported' to other countries. Bucht (2002) illustrates the importance of mutual international inspiration and 'learning', as Swedish urban park planning was, on its turn, influenced from England and Germany. This influence related, for example, to the 'moral' reasons for establishing public green, that is to offer 'proper' and healthy recreation to the lower classes.

Over time, prime examples of what were considered successful city forests were appreciated and visited by decision-makers, designers and green space managers, offering inspiration for city forest and park development in other countries. Those who designed the Kralingse Bos of Rotterdam during the early 20th century, for example, were inspired by the German Volkspark concept and the American Park System movement (Andela, 1996). They attempted to combine new ideas about 'free nature' and more natural parks with those of promoting active outdoor recreation. The planners and designers of the Amsterdamse Bos visited landscape parks in England and the Volkspark in Hamburg during the 1920s and 1930s

(Bregman, 1991; Blom, 2005). Many years later, those responsible for the English Community Forests visited areas such as the Amsterdamse Bos to study the characteristics of a successful city forest landscape (C. Davies, 2001, personal communication). Dutch landscape architects designing new woodlands in their country during the 1990s also obtained at least part of their inspiration from abroad. A report analysed Epping Forest near London, the Berlin Grunewald and the Forêt de Saint-Germain near Paris as prime examples of successful city forests. The strengths and weaknesses of these three city forests were analysed for the benefit of the design of new, Dutch city forests (Slabbers et al., 1993).

Thus, over time, well-known city forests such as the Amsterdamse Bos became reference landscapes for the establishment and management of (city) forests across Europe. Epping Forest is another example of such 'reference city forests'. According to its management plan, it should become a nationally-recognised centre for ecological and nature conservation research (City of London Corporation, 1998). Fontainebleau is yet another example, although it is perhaps more an 'urbanised' forest than a city forest, given its distance of more than 50 km from Paris. Not only was this forest a test case for nature conservation in France and – in the eyes of some – a 'national museum' (Kalaora, 1981), the French State Forest Service has also regarded Fontainebleau as a model for forest management in France and abroad (Trébucq, 1995). This is reflected, for example, by its central role in an international project which has the main aim to reconcile the needs of conservation of forests with those of recreation (PROGRESS, 2007).

Demonstrating Forestry to the Public

We have seen how city forests have acted as testing grounds, facilitating the development of modern forestry and communication between experts. They have also reflected international developments and fashions, for example in forest and landscape design and management. Yet another role of city forests has been their role in communication between foresters and the public. Where better explain what forestry is all about than in forests close to where most people live?

Communication has not always been a very strong side of forestry. Traditionally, foresters have often communicated their management decisions and actions through rather technical reports and plans. Given a mandate by forest owners, they could do their work mostly in rather remote areas, not controlled by the urban populace. However, this has gradually changed, together with changing societal demands for multiple forest goods and services and with increasing urbanisation. Another driving force leading to a call for more interaction between forestry professionals and society is the democratisation process and people's desire to be involved in decision-making that affects their living environment (Kennedy et al., 1998; Konijnendijk, 2000; Van Herzele et al., 2005). At a 1962 conference on Suburban Forest and Ecology held in Connecticut, USA, this challenge was already touched upon (Waggoner & Ovington, 1962). At the conference, Forestry and other experts stressed the need for more communication with urban dwellers. Foresters also needed to better understand the natural environment in which a growing number of Connecticut citizens were building a new, urban way of life.

Research has shown that a proper communication of options with other experts as well as with the public at large requires new methods (e.g. von Gadow, 2002; Tyrväinen et al., 2006, 2007; Janse & Konijnendijk, 2007). The forests of Berlin and the Zoniënwoud of Brussels, for example, can be visited virtually through a forest walk on the Internet, and different forest management methods are explained online (Berliner Forsten, 2006; Région Wallonne et al., 2007). However, there is evidence that management options for urban woodlands need to be demonstrated and talked about in 'real life', that is in the forest, to become more understandable and easier to discuss (e.g. Tyrväinen et al., 2006).

As mentioned, forests close to cities offer excellent opportunities for foresters to demonstrate their skills and the importance of forests to urban populations that are gradually losing contact with nature. The Belvoir Park Forest near Belfast, Northern Ireland, has been described as a 'microcosm of forestry in Northern Ireland' (Northern Ireland Forest Service, 1987). The park, which was opened to the public in 1961 and in which commercial forestry is still practised, offers excellent opportunities for local foresters to show what forestry is all about (Simon, 2005).

Communication opportunities in city forestry can be enhanced even further when special demonstration forests are established (von Gadow, 2002; Tyrväinen et al., 2006). Demonstration forests can combine several functions, as they provide a platform for discussions about management options with other professionals, interest groups and the public at large. In this way, forest management becomes much more tangible and transparent, and conflicts might be avoided. Importantly, demonstration forests can also help show the multiple benefits of forests to society and the role of forestry in providing these. In line with the traditional role of city forests as testing grounds, which was described earlier in this chapter, demonstration forests can be used to develop innovative management approaches, for example in response to urban pressures. Demonstration forests can have different scopes and structures, depending on the local context and local needs. For optimal benefit in terms of communicating with the public, however, they should be situated close to where many people live. By means of good design, a wide range of forest types and management alternatives can be demonstrated within a relatively small area. Good educational interpretation, recreational facilities and attractive events should be provided to enhance their use. This is crucial, as demonstration forests will only function well if intensively used by both foresters as well the public.

In a way, arboretums were amongst the first 'demonstration forests', as they were often open to the public and showed different tree species and how these grew. The Tervuren arboretum mentioned before played an important role not only in testing different combinations of tree species, but also in showing visitors what forests across the world looked like. However, it has not really met public demands for sufficient information and recreational facilities (van der Ben, 2000).

During the past two decades or so, several demonstration forests have been created across the globe. Some of these are large-scale and not specifically targeted towards urban audiences, such as the Canadian Model Forest Network. The Polish national programme of 'Forest Promotional Complexes' has had a similar scope as the Canadian programme, i.e. to develop and demonstrate innovative forestry approaches. However, it has placed more emphasis on the recreational and educational needs of urban populations. A number of the 'promotional sites' with an extensive educational and recreational infrastructure are therefore situated close to urban centres (Konijnendijk, 1999).

Also elsewhere are demonstration forests are situated close to where most people live. Several of these forests are owned or managed by universities and also used as education and testing areas. During the 1990s, forest researchers with the Swedish University of Agricultural Sciences established a demonstration forest called the 'Urban Forest of the Future' (Framtidens Skog) to demonstrate new principles in the silvicultural treatment of young urban woodlands (Rydberg & Falck, 1998). The target audiences are professionals but in particular also to the general public. The 12 different compartments in the forest are open to visitors and visitor preferences are of primary importance for guiding management.

Other urban woodland owners have also set up demonstration and reference areas. Demonstration areas are part of the silvicultural management strategy applied in the municipal forests of the German cities of Luebeck and Goettingen, while the city of Arnhem, The Netherlands, has set aside part of its forests to demonstrate different silvicultural options and systems to the public (Konijnendijk, 1999). Also in The Netherlands, a demonstration forest was established by the city of Ede during the 1980s, in response to public demand for better information about forest management (Lub, 2000). The Flevohout woodland near the town of Lelystad was set up a few years ago to demonstrate (and test) various management systems aimed at producing European quality timber. The Flevohout has had the objective to act as 'unique meeting point' between foresters and all parts of the forest-wood chain. Additionally, public access and information are prioritised by those responsible for the Flevohout, the Robinia Foundation (which promotes the production and use of European quality timber) and the Dutch State Forest Service (Jongheid, 2000). In St. Petersburg, Russia, Danish experts assisted local foresters with the development of the Toxovo Demonstration Forest in the 142,000 ha forest green belt (Nilsson et al., 2007).

Communicating with the public is not only about demonstrating. Ways have to be found to better capture and incorporate local knowledge and expertise. In an interactive planning project in the Ronneby Brunn woodland park of Ronneby, Sweden, Gustavsson et al. (2004) stressed the role of so-called 'connoisseurs', people with a deep local knowledge of a certain aspect of the forest, its use or management. Connoisseurs are often 'non-professionals', such as amateur ornithologists, leaders of local scouting clubs or handicapped users of a forest. Their knowledge and a sound understanding of their preferences is crucial for developing 'good', appreciated city forests. The importance of local knowledge in the management of city forests is also stressed by Jones and Cloke (2002), who offer several examples of how local people develop special, close relationships with woodland. If these links and special uses are not recognised by forest planners and managers, conflicts can easily occur (see Chapter 12 'A Forest of Conflict').

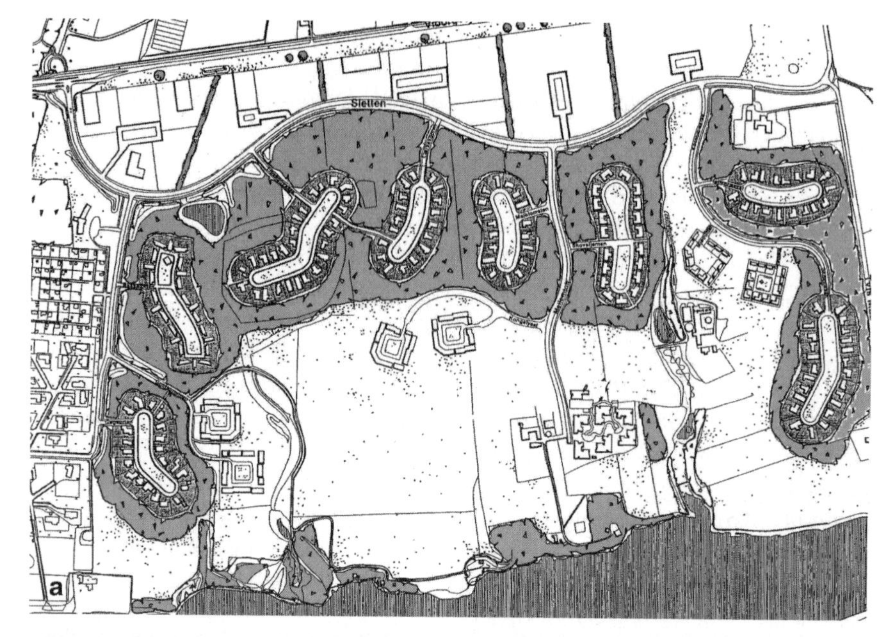

Fig. 10.2 The Sletten Landscape Laboratory in Holstebro, Denmark, comprises 'islands' of new housing situated within different types of woodland **(a)**. Around the houses, in many cases a smooth transition has developed from (private) garden to (public) woodland **(b)** (Map by Holstebro municipality, photo by Anders B. Nielsen)

In Denmark, Sweden and elsewhere, the topic of demonstration forests as true means of two-way communication amongst experts and between experts and the public has been taken one step further by developing 'landscape laboratories'. These are experimental areas the size of a local landscape which typically include woodland as well as other landscape elements (Gustavsson, 2002; Tyrväinen et al., 2006). By showing the importance of long-term management strategies and stressing the use of creative management to differentiate the landscape, the landscape laboratories also provide 'living references'. These landscape laboratories were initiated by the Department of Landscape Planning Alnarp, Swedish University of Agricultural Sciences, under the leadership of Roland Gustavsson. One example, the Swedish Snogeholm Landscape Laboratory, was set up in 1994 and is focused on afforestation and forestry practice. It encompasses over 60 different woodland interior and edge character types, many of which have not been applied in forestry. More recently, a laboratory was also established at the town of Holstebro, Denmark. This Sletten Landscape Laboratory is the most 'urban' of the present landscape laboratories (Fig. 10.2). It was developed together with three new housing areas, providing a 'green framework' for the new houses. The landscape laboratory provides a 'green linkage' between city and countryside. Another landscape laboratory will be developed in the Emscher Landschaftspark (Emscher Landscape Park), situated in the German Ruhr area (Lohrberg, 2007). This laboratory will be established at a site where factory premises and mining dumps of the colliery are to be transformed. The new woodland area should help the re-development process by strengthening the area's unique character and image through biomass production and recreational use. It also offers possibilities for improving the quality of life in the neighbouring quarter of Dinslaken, an area which presently faces many social problems.

The development of landscape laboratories shows that the educational roles of city forests are being expanded. Their traditional use as scientific and professional testing areas is being combined with a focus on 'joint learning' of experts and the public. This also links up to another educational role of city forests, namely that of settings for nature education involving especially children and youths.

City Forests, Nature Education and Children

Today's urban dweller may have many wrong ideas about forests and nature, simply because of lack of contact with nature and natural processes. Many people are no longer 'connected' to forestry and rural life and many of their opinions about nature and forests are shaped by the news media (Otto, 1998). Nature educations efforts attempt to 're-connect' people to nature and to explain them about the importance and workings of nature. City forests play an important role in nature education – a role which goes beyond providing a setting for discussions about forest management.

Many city forests host educational facilities such as visitors' centres (Fig. 10.3), nature trails, and information panels. Nature education plays an important role in the forests of Berlin. There are various visitors' centres, including the 'Oekozentrum'

Fig. 10.3 The Watermolen (Water Mill) visitor centre at the Sonsbeek woodland park of Arnhem, The Netherlands, is run by a non-governmental organisation (Photo by Sophia Molendijk)

('Eco-centre') in the Grunewald, which is run by a special foundation which attempts to make people aware of nature and environment (Konijnendijk, 1999). The first 'forest educational path' in Austria was established in 1965, in the Lainzer Tiergarten near Vienna (Ballik, 1993). Djurgården in Stockholm hosts a wide range of educational facilities, including an open air museum (Nolin, 2006). The managers of many city forests engage themselves in guided walks and presentations.

Children and youths are a very important target group for nature education, not in the least because studies indicate that outdoor play is important for children's developments (e.g. Taylor & Kuo, 2006; O'Brien & Murray, 2007). The importance of outdoor learning and play, and the 'rural' needs of the urban child were also recognised by Friedrich Froebel, the founder of the kindergarten concept (Hiss, 1990). Wilderness experiences and outdoor challenge programmes seem especially important, as these have been found to promote self-esteem and sense of self amongst children and youths (Taylor & Kuo, 2006). This is illustrated by a study in Britain, which found 'wild adventure space' to strengthen the personal development, as well as social and community skills of youths between 12 and 18 years old (Thompson et al., 2006). Studies in North America indicate that children who play in natural environments undertake more creative, diverse and imaginative play (Taylor & Kuo, 2006; O'Brien & Murray, 2007).

Spencer and Woolley (2000) mention the importance of fostering participation of children in a city's social life. Too often children are unacknowledged outsiders in the planning and management of urban areas. Place attachment is important in developing a child's personal identity. This calls for more direct involvement of children in the nearby natural environment as well. Woods and trees are an important part of this nearby natural environment. Interviews with kids from Argentina, Australia, Mexico and Poland compiled by renown architect and author Kevin Lynch all demonstrated an outspoken and seemingly universal hunger for trees (cited by Hiss, 1990).

But are children using nature and their local city forests? In a recent interview (Talking with the author, 2007), American author Richard Louv, who wrote the bestseller 'Last Child in the Woods' (Louv, 2006), speaks of 'nature-deficit disorder', which he sees as handy way to identify the disconnection between children and nature and the implications of that on health (of the child as well as the earth). Children today are increasingly sedentary and it is not as obvious as in the past that children spend a lot of their time playing outdoors. A survey amongst Danish 10–14 year olds on attitudes towards nature confirmed that there is reason for concern, even although nature still features prominently among children's preferred activities – 15% of the respondents even preferred 'studying nature' (Danmarks Naturfredningsforening & TNS Gallup, 2004).

Denmark and other Nordic countries do take environmental education and keeping children in close contact with nature seriously. So-called 'forest schools' are run by municipalities and the state forest service, enabling children and youths to learn and be active in nature for a day, a week or even longer. Forest schools offer the opportunity to gain access and become familiar with woodland on regular basis, while learning theoretical and practical skills. Other countries have also introduced the forest school concept. The first forest school in Austria was opened in the Wienerwald during May 1998 (S. Redl, 1998, personal communication), while Berlin's forest service runs six forest schools, based on the example of the 'Sihlwaldschule' situated near Zurich (Berliner Forsten, 2006). Many forest schools are booked for months or even years ahead. O'Brien and Murray (2007) found that forest schools in England and Wales offer many benefits, such as a positive impact on children in terms of confidence, social skills, language and communication, motivation and concentration, physical skills and knowledge and understanding.

Children can also be more actively involved in the management of their local environment and local forest, as promoted by Spencer and Woolley (2000). Moigneu (2001) describes the case of the Forêt de Sénart, a 140 ha woodland in the outskirts of Paris. This state forest had experienced serious abuse by people from surrounding residential areas characterised by high unemployment, crime and violence. In response to this, a 'Junior Foresters Education Programme' was then set up by ONF, the French State Forest Service, in collaboration with local schools. A large part of the children from the multi-ethnic residential areas was enrolled in the programme, which involves various forestry activities, from cutting trees to establishing recreational facilities, carried out

Fig. 10.4 Children are of primary importance for environmental education. City forests provide a good, close-to-home setting for educational activities, as shown for this example from Israel (Photo by Moshe Shaler)

together with professional foresters and nature interpreters. Children who 'graduate' from the programme receive a diploma and the title of 'Junior Forester'. The programme has been very successful, as it has resulted in more social control, less vandalism and better relationships between foresters and the local community. Allegedly, the participating children have drastically improved their school grades in biology as well.

As illustrated before, regular visits to city forests and nearby nature by kindergartens and schools are important (Fig. 10.4). Kardell (1998) noted an increased use of Norra Djurgården in Stockholm by nursery schools from the 1970s to the 1990s. The Swedish Association for the Promotion of Outdoor Life played an important role in this positive trend. A study in Denmark by Holm (2000) confirmed these findings, identifying an increased use of urban green areas by day-care centres and other institutions. However, school use of green space is under pressure, as it requires time, resources and raises concerns with parents, for example concerning child safety. Schantz and Silvander (2004) mention that since 1994, visits to forests and nature areas by schools in Sweden have in fact been decreasing, as there no longer are national rules for a fixed number of 'outdoor' days for schools. This is a worrying development.

Tomorrow's Forest of Learning

City forests have played and will continue to play an important role as Forests of Learning. First of all, the role of city forests in the advancement of forestry at large needs to be recognised and developed. As mentioned by Ronge (1998, p. 37): "It is possible that town foresters are – by tradition and because they normally have more direct contact to the population – not as much in conflict with ordinary people as the ordinary foresters of large state or private forests." City forestry has indeed been at the forefront of innovations in forestry (Krott, 1998, 2004; Konijnendijk, 2003). Communication with the public is often more advanced in city forestry than in other types of forestry. This is probably due to the simple fact that city foresters have always had to deal with the many demands of a large urban population and local politics. As discussed above, education has often been a natural task of city foresters, from guided walks to school visits and even running forest schools. Moreover, city forestry has had to work closely together with other fields, for example to conserve and develop a city's overall green space resource.

City forests continue to be testing grounds for forestry and landscape management. One example of a 'city forestry innovation' is described by Kowarik (2005), who describes 'wild urban forests', forests that spontaneously develop on former industrial sites, as a new type of forests. More about these wild urban forests is written in Chapter 8 'The Wild Side of Town' and Chapter 13 'A Forest for the Future'.

The landscape laboratory concept is being developed further, often within the context of city forestry. Landscape laboratories also offer an interesting tool for one of the major challenges facing city forestry today: how to plan, design and manage new urban woodland. Across Europe, the woodland cover is increasing. Afforestation is often taking place close to cities and towns (e.g. Nielsen & Jensen, 2007). Experts doubt whether sufficient attention is being given to visual aspects and design of these new woodlands (e.g. Gustavsson, 2002; Wiegersma & Olsen, 2004; Nielsen & Jensen, 2007). Gustavsson (2002) stresses the need for approaches that take an integral look of planning, design and management of urban woodland. Management, he states, is an integral, creative and long-term development process. Design does not stop when the forest has been established, as the woodland has to be given an identity of its own. This can only be done in close collaboration with local people, who will be the main users of the new city forest. A particular problem is also related with the forests that were planted several decades ago. These 'teenager forests' will need to be carefully developed and shaped, in order not to end up with forests that all look the same and are not very attractive for recreational and other uses. Nielsen and Jensen (2007) offer some valuable tools for improving the visual qualities of urban woodland. The use of succession and variation in tree species, age and spacing, for example, can enhance the experience of diversity and naturalness, even in young woodland.

One of the future 'educational roles' of city forests as testing ground may relate to the huge challenges connected with global warming. Theil (2006) mentions how megacities create a mosaic of habitats and microclimates, from pond-filled gardens to industrial brownfield sites. Cities are turning into vast laboratories for studying animal behaviour and evolution. Moreover, as cities have often been several degrees warmer than surrounding countryside due to a phenomenon known as the urban heat island effect (e.g. Akbari et al., 1992), valuable experience has been gained with growing trees and vegetation that are better adapted to heat stress. City forests can also help raise awareness about climate change issues, making (often symbolic) contributions to mitigating climate change. In The Netherlands, for example, a national campaign promotes the establishment of so-called 'climate woods' ('Klimaatbosjes') consisting of at least three walnut trees (*Juglans regia* L.). Many schools and businesses have already established their own climate wood (Zuid-Hollands Landschap, 2006; Nederland Klimaatneutraal, 2007).

As described, communication with the public is an important aspect of the Forest of Learning. To promote the importance and use of city forest in today's information society, where people are continuously bombarded with information and where competition for people's time is fiercer than ever, innovative ways of teaching and communication will have to be developed. The example of the new Pritzker Family Children's Zoo at Chicago's Lincoln Park, USA, shows that innovative nature education does not have to be high-tech. The design team for the zoo had the ambition to create an 'immersive woodland experience', with blurred boundaries between animal habitat and visitor space. The zoo only hosts native animal species and resembles a natural landscape. The need to offer various opportunities for learning guided the design of the zoo (Dennis, 2006).

City forests and other green areas near cities can also be used to make links between urban dwellers and historical uses of the land (e.g. Dietvorst, 1995). Many open air museums, for example, are situated in city forests, showing how life was in historical settlements and raising awareness about traditional agriculture, horticulture and forestry.

As discussed, the role of city forests as 'wilder' nearby nature (see Chapter 8 'The Wild Side of Town') is crucial for future environmental education efforts and particularly for those aimed at children and youths. City forests can be 'urban wildscapes', areas that are not too intensively managed and offer children opportunities for play, discovery and adventure. In this way they learn about nature as well as about themselves and their relations to others. But barriers will have to be overcome, such as society's fears of letting children play in the woods, as parents fear that their kids will get hurt or lost. In a recent European project called 'NeighbourWoods', a local school in Helsingborg, Sweden, agreed to work together with foresters, nature interpreters and scientists, taking part of its education programme to a local city wood. Young children were involved in forest management activities and granted freedom to create their own places in the woodland, the Filbornaskogen. The researchers who studied the process noted that children developed their creativity, understanding of the forest, as well as relations with each other (Gunnarsson & Palenius, 2004). However, when a movie about the project was shown at an international conference, participants from several other European countries reacted strongly when they saw young children walking through the forest with saws and other tools in their hands. This would never be possible at home, people said, because of parental concerns and the many rules and regulations concerning liability. This shows that there is still a long way to go in using city forests as 'wildscapes' in environmental education efforts.

Chapter 11
The Social Forest

The previous chapters have looked at the cultural landscape of city forests from different perspectives. This chapter highlights another important role of city forests, namely the contributions these forests (can) make to the development and identity of local urban communities. Trees and forests have been important in 'place making' by local communities, providing local identity, a meeting place and a link between people and their local environment.

The first section of this chapter looks at the evolution of traditional wooded commons into city forests, a topic also discussed in previous chapters. Focus here will be on the social and community identity aspects of wooded commons. Most commons were taken by the state or privatised. However, from these areas, city forests emerged as a new kind of 'forest for all', as public spaces designed primarily to offer opportunities for recreation. These social forests became characterised by high levels of recreational use, especially from nearby residential areas, as described in the second section. The third section of this chapter focuses on city forests as 'social stages', places where urban dwellers can meet and socialise, places which help (re)confirm community ties. Finally, the chapters looks specifically at the role of city forests in place making, community building and providing identity to cities. It is stressed that socially-inclusive planning and management of city forests are crucial for developing the Social Forest.

From Wooded Commons to Forests for All

Previous chapters have described general developments in society that are of relevance for city forestry, such as changing power and social relations, changing nature views and the emergence of a leisure society. City forests, their ownership and use are a mirror of these societal developments.

City forestry has a highly democratic and public origin, as many of the areas that would later become city forests emerged from common lands, which were jointly managed and used by local communities. Through customary rights that existed throughout Europe, peasants could use forests for certain purposes, including grazing and litter removal for fodder. These rights were considered very important by the

C.C. Konijnendijk, *The Forest and the City: The Cultural Landscape of Urban Woodland*

poor (Holscher, 1973; Westoby, 1989; Jeanrenaud, 2001). In many European countries, most villages had their commons, for example in the form of a village green (Olwig, 1996). Town meetings were held to regulate the use of these commons. They were used for grazing, but perhaps more importantly as a venue for community events and festivals. Olwig (1996) argues that the commons thus did not only play an important economic role, they also contributed to strengthening community bonds and identity.

As described in Chapter 5 'The Forest of Power', gradually wooded and other commons were claimed by rulers, the state and private owners, primarily to restrict their use to only few activities (hunting, later timber) and groups (i.e. the elite), and also to fight overuse. Forest areas of traditionally common use, also near cities and towns, were fenced off. Municipalities started to become more active in expanding their forest property. Obviously the desire to extract timber and other forest products from the woods was an important motivation. But recreation also came into the picture, as earlier chapters such as Chapter 6 'The Great Escape' have shown.

Urban growth and especially the Industrialisation led to deteriorating conditions in many European cities. The rapidly growing population had to turn somewhere to spend its leisure time. Historically, the landless workers of the larger cities had few places to go for recreation and relaxation. Industrialisation and the building of large-scale factories now led to the transformation of many of the open spaces that still remained. As a result, municipal forests and the countryside ringing the cities offered the only available sites that offered escape from the urban environment (Holscher, 1973; see Chapter 6 'The Great Escape'). Thus city forests, together with other areas such as cemeteries, played an important role as 'recreational escapes' even before public parks were developed more extensively. In many European countries, particularly in the northern and central part of the continent, a traditional everyman's right ensured public access to forest. People were (and still are, in many cases) free to go wherever they want, as long as they did not cause any damage to the forest, or, in most cases, refrained from taking timber (Jensen et al., 1995).

Many cities did harbour large green areas other than commons, but these were mostly owned by the aristocracy and by the bourgeoisie, and their use was generally restricted. So-called 'public use' (for recreation) initially referred only to the use by certain 'well-behaved' segments of society. In St. Petersburg, Russia, for example, a clear division was made between the elite and a large class of 'country bumkins' when it came to the recreational use of the imperial parks and gardens (Anan'ich & Kobak, 2006). Most ordinary people were barred from the parks, although stories indicate that certain 'lowly' elements such as prostitutes were allowed in.

Many of the royal gardens and forests in Britain were opened before their counterparts on the continent and even before the Industrialisation. Hyde Park and St. James's Park were opened for (at least limited) public use as early as the 17th century (Hennebo, 1979, p. 74). London's first public 'pleasure garden', Vauxhall Gardens, was opened in 1661, although people had to pay an entrance fee to get in (Reeder, 2006b). Other parks were incidentally opened for festivals. Not everyone was happy with this, however. During the 18th century, requests were still being made – undoubtedly by the elite – to limit the public use of the royal parks (Hennebo, 1979). According to Reeder (2006b), the English Victorian elite had a strong distaste for the antics of day-trippers in Epping Forest, which had been made more accessible to the working class population of London's East End by rail. People also spoke their distaste of the 'frolicking' taking place on Hampstead Health and stated that 'the people' seemed to have forgotten how to amuse themselves properly. After the opening of its own railway station in 1860, the Heath was attracting tens of thousands of visitors on sunny days, becoming perhaps the most popular open space of London by the end of the century (A history of, 1989). City forests and other more natural areas thus seem to have offered more freedom and less 'social control' to the working classes than inner-city municipal parks.

In France, the 1789 Revolution paved the way for more public green space. Access to formerly private parks and gardens was enhanced and new green spaces were created. This trend could also be noted in other parts of Europe. Among the areas opened up to the public for recreational use were the former hunting forests and parks in and around the cities. In the case of France, the hunting forests near Paris soon became popular destinations for citizens, who for example admired the deer in the Bois de Boulogne. Visitor numbers peaked on Sundays and during public holidays (Hennebo, 1979). Deer were also a prime attraction in the Jægersborg Dyrehave royal hunting park north of Copenhagen. After its opening to the public in 1756, the park became one of the most popular recreation areas for the inhabitants of Copenhagen (Møller, 1990). Berlin's Tiergarten was transformed into a park-like space allocated for citizen recreation (Nehring, 1979) at the end of the 18th century.

Apart from opening of former royal (hunting) domains to the public, new green spaces were also established for the benefit of the general public. The 19th century German 'Volksparks' were open to all classes and thus were real social green spaces (Hennebo & Hoffmann, 1963). The Volksparks or 'People's Parks' were characterised by clean air, freedom and beauty of sky and landscape for public enjoyment. Activities such as sports and games, and just being outside, were initially considered more important than the aesthetic considerations.

These principles were also followed elsewhere in Europe as more municipal parks were established or developed from existing green spaces. The continuing transformation of hunting domains into public use, for example, often had the principles of the Volkspark in mind. Contributing to the image of the royal residencies based in the domains remained important, but recreation and concerns for public health had come to the fore. The Grunewald in Berlin underwent such a transformation. Following a decision by the German Emperor, it lost its function as a court hunting domain at the beginning of the 20th century and became a Volkspark owned by the regional government of Berlin. A special decree called the 'Dauerwaldvertrag' (permanent forest agreement) was drawn up to ensure that the forest would remain in public use. The fondness of the people of the Grunewald was demonstrated by the 30,000 signatures collected against plans to carry out large-scale timber cutting in the forest (Cornelius, 1995). Areas such as Grunewald and Tiergarten were needed, as Berlin offered very little outdoor recreational opportunities for its population.

Similar developments towards more 'social' forests and green space were seen elsewhere in Europe. Municipal authorities had gained more power and administrative capacity. In Stockholm, Sweden, city authorities took over the management of public parks and green space in 1869. 'Social' parks were established during the period 1880–1930 (Nilsson, 2006a), and these parks had close links to the German Volkspark movement. The new parks were to supplement the existing (mostly royal) green areas. The royal woodland park of Djurgården continued to play a crucial role for recreation (Nolin, 2006), remaining as one of the most important open spaces in town despite the new parks being commissioned. The level of facilities in Djurgården was raised, for example to provide ample opportunities for sports and games. Remaining natural areas, such as the Hellasreservat near the new urban area of Hammarbyhöjden, were bought by local government and reserved for open-air recreation purposes (Nolin, 2006). Helsinki in Finland was in the fortunate situation that nature had remained readily available for all its citizens. Debates about inner-city green space had up to the late 19th century been driven mostly by the city's elite. However, public debate did emerge about the lack of green spaces in working class neighbourhoods. Interestingly, city parks were still heavily regulated by special officials as late as the 1960s (Clark & Hietala, 2006).

Apart from the Volksparks and the general development of public parks, social and outdoor leisure opportunities were also improved in other ways. Allotment gardens were created across Europe, as where greener suburbs and New Towns. In relation to allotment gardens in Helsinki, Lento (2006, p. 198) writes: "Reformers saw them as offering the working class a hobby that would introduce them to the way of life in the countryside, as well as strengthening their connection with the 'fatherland'." Thus, community and nation building and pride, given more attention later in this chapter, were a clear objective.

The 20th century brought further urbanisation and increases in welfare and leisure time. City forests were particularly needed to offer recreational opportunities to the growing urban population that now had more time to spend, as described in Chapter 6 'The Great Escape'. Access and facilities also had to be improved for this purpose, and city forests had to be turned into real 'social forests'. In The Netherlands, the establishment of a new recreational area south of Amsterdam started in 1934, based on a 1928 plan. This Amsterdamse Bos was conceived as a real 'social forest'. The responsible 'Commission for the Forest Plan' (Bosschplan) stressed that the population was to consider this Amsterdamse Bos as its own, without any restriction. The Commission stated that, unlike in other forests in the western part of the country, the population would not have to feel welcome on a temporary notice only (Balk, 1979). The Amsterdamse Bos later provided, together with the Kralingse Bos of Rotterdam, an important example for massive recreational developments in The Netherlands in the post-World War II period, as set out in the 1963 national policy note 'Recreation spaces in The Netherlands, (Konijnendijk, 1994; Blom, 2005).

Social Forests for Mass Use

As also described in Chapter 6 'The Great Escape', the recreational use of city forests grew rapidly during the 20th century with further increases in leisure time and mobility (e.g. Fairbrother, 1972). Improved access and infrastructure, and the emergence of a strong consensus on the 'healthy' type of recreation these green areas offered, also played important roles. Recreation in the Berlin forests, for example, had increased rapidly and was turning into mass use (Cornelius, 1995).

Today, large proportions of city populations are using the local city forests, turning them into real 'community forests'. A telephone survey by Bürg et al. (1999) in Vienna, for example, showed that between 66% and 80% of the population visited the nearby Wienerwald (Viennese Forest) during the period 1993–1997. No less than 28% of the Viennese visited the Wienerwald every day. A similar level of popularity for city forests was found for Hamburg, where in 1994, 80% of the population used the local forests at least once a year (Elsasser, 1994). In Berlin, 36% of the respondents in the study – deemed representative for the entire city population – visit a city forest every weekend, and 3% daily. Twenty-three percent of the respondents said they visit the forests once a month (Meierjürgen, 1995).

Compared with other forests and nature areas, city forests are thus real 'social forests', used by a high share of local residents (Fig. 11.1). This also results in

Fig. 11.1 City forests, such as the Sonsbeek park of Anrhem, are real social forests, being used by local residents for a wide range of recreational activities (Photo by Sophia Molendijk)

very high visitor numbers. Jaatinen (1973) mentions that, as early as 1973, about 2 million visits per year were paid to the Keskuspuisto ('central park') city forest in Helsinki. A study of urban woodlands in major European cities showed that forests situated in or bordering cities typically attract 2,000 or more visits per hectare per year, which is much higher than the visitation rate for 'regular' forests (Konijnendijk, 1999).

These findings are confirmed by a comprehensive national survey of recreational use of forests, nature and other open areas in Denmark (Jensen, 2003). This survey showed that forests and green areas in the Greater Copenhagen area have some of the highest visitation numbers. The 'top scorer' is Jægersborg Dyrehave & Hegn, north of Copenhagen, which attracts close to 7.5 million visits per year, a stunning average of 4,460 visits per hectare per year. In The Netherlands, nice sunny Sundays during the 1960s could draw more than 100,000 people to the 900 ha Amsterdamse Bos for sunbathing, picnicking and swimming (Balk, 1979; Bregman, 1991). By the end of the 1990s, more than 4.5 million people visited the Bos annually (Van den Ham, 1997). For European Russia, it has been estimated that – depending on the season and time of the week – nearly one-third of the population at any given time may be utilising the (mostly peri-urban) forests for recreation (Barr & Braden, 1988). The fact that many Russians spend large parts of their summer in their 'dacha' or summer cottage obviously contributes to this.

Often, as also mentioned Chapter 6 'The Great Escape', activities such as walking, cycling and observing nature are most popular in city forests. But more 'social' and massive forms of recreation also take place in these forests, as the previous examples of Jægersborg Dyrehave and the Amsterdamse Bos make clear. Many city forests host facilities for mass recreation activities, such as swimming, picnicking, attending special events such as concerts (Fig. 11.2), or visiting special attractions (amusement parks, zoological gardens) in high numbers. This is not a new phenomenon, as fairs, such as the annual Dunlop Fair in Epping Forest (Green, 1996), brought out the masses to city forests as early as during the 18th and 19th century. During modern times, the Malieveld, a large open area bordering the Haagse Bos, has hosted many political, entertainment and other events. In 1988, 86,000 people attended a Rolling Stones rock concert (Anema, 1999). In Berlin, Freibad Wannsee situated in a city forest had become the largest lake pool of Europe by 1907 (Cornelius, 1995). Sports organisations have also had a large influence in the establishment of a recreational infrastructure in city forests (e.g. Lento, 2006 in the case of Helsinki). Norra Djurgården in Stockholm has gradually become an area for large-scale commercial sporting competitions, attracting as many as 15,000 participants (Schantz, 2006).

Commercial developers and entrepreneurs have benefited from the mass recreational use of city forests and city parks. Semenov (2006) mentions how the 'leisure boom' led to the involvement of commercial developers in the green spaces of St. Petersburg. Special amusement parks were established around 1900. Ponds in the historical Tauride gardens were used for commercial leisure angling. Today, commercial amusement parks such as Grönna Lund in Stockholm's Djurgården and the previously mentioned Bakken in Jægersborg-Dyrehave near Copenhagen are a

Fig. 11.2 City forests are social stages where people interact. They also provide an actual stage for events such as concerts which can attract large audiences. This example shows a concert in the Sonsbeek woodland park of Arnhem, The Netherlands (Photo by Sophia Molendijk)

main attraction for many visitors. They have also attracted people to singing and dancing events, offering them a mix of tradition and community (e.g. Nilsson, 2006a). Other paid attractions in city forests include zoological gardens, which are a very common element. Examples of city forests and parks with a zoological garden are Dublin's Phoenix Park (Nolan, 2006), the wooded part of Vienna's Schoenbrunn Garden and the Berg en Bos city forest of Apeldoorn, The Netherlands (e.g. Konijnendijk, 1999).

The social role of city forests and their mass recreational use have been supported by the policies of post-war local and national governments. These bodies did not only improve the accessibility and facilities of their own, public urban forests, but also stimulated them in private forest areas. In many European countries, access to forests has always been good due to the traditional everyman's right. In many Western European countries, however, public access to especially private forests has not always been that obvious. Currently in The Netherlands, the government compensates forest owners for opening up their forests to the public. In Denmark, public access to private forests for walking and bicycling is allowed during daylight hours, and only where there are access roads to the area (Konijnendijk, 1999).

Even areas further away from the city have become 'city forests', given that their use is dominated by citizens from one specific city. The forest of Fontainebleau and its links to Paris, frequently referred to in this book, is an example of this. However, distance to the forest is an important factor, and walking time has clearly been identified as the single most important precondition for use of urban green space in general (e.g. Farjon et al., 1997; Van Herzele & Wiedemann, 2003; Tyrväinen et al., 2005). This also means that most city forests are particularly used by local people (e.g. Hörnsten, 2000). Thus green spaces such as city forests should be near by where people live, in order to be used regularly and to be seen as part of daily life. 'Near' in this sense mostly means within easy walking (or sometimes cycling) distance, i.e. 300–500 m.

A comparative study of Nordic cities mentioned that 85% of the population of Helsinki and 95% of the residents of Oslo has 300 m or less to travel to the nearest green area of at least 0.2 ha in size (Silfverberg et al., 2003). Unfortunately, not all European city dwellers are as well off. Twenty-six out of the 30 biggest cities in The Netherlands, for example, included areas without proper access to public green space (Gerritsen et al., 2004).

Studies have also specifically looked at access to urban woodland. Research of woodland use in the British town of Redditch showed that over 70% of the users had less than a five-minute walk to the woodland which they used (Coles & Bussey, 2000). A recent survey by the Woodland Trust (Cooper & Collinson, 2006; Woods for People, 2007) showed that only 46% of all woodland in England is accessible. According to a national woodland access standard, no person should live more than 500 m from at least one woodland area of 2 ha or more, while a woodland area of at least 20 ha should be situated within 4 km. According to this standard, 55% of the English population has access to large woodlands (of at least 20 ha), but only 10% to nearby small woodland areas (i.e. of 2 ha or more). In Antwerp, Belgium, only 15% of the population has a city forest or urban woodland within reasonable proximity (Van Herzele & Wiedemann, 2003). As described above, city forests play a key role in developing city societies. Therefore, if not every citizen has good access to the green structure, one could speak of 'social fragmentation' of the landscape. Chapter 5 'The Forest of Power' looked at this issue from the perspective of environmental justice.

In spite of limited availability of city forests in some cities and for some segments of the population, city forests have gradually become truly 'social forests'. They are frequently used by the majority of city dwellers and, unlike other forests, they sometimes are a setting of mass recreational use. When considering these 'social city forests' today, the question could be asked if some sort of common rights still exist. In some cases they do, be it on a much smaller scale. In the case of Epping Forest, for example, there are still commoners' rights, but only a handful of those who quality actually graze their cattle in the forest (The Corporation of London, 1993).

City Forests as Social Stages

As seen, today's city forests are mostly accessible to the public and attract a large number of leisure seekers, especially from the adjoining or nearest city. Probably more than other forests used for recreation, however, city forests play a role as elements of public open space. They are often an integral part of city life in its many aspects.

Gadet (1992), who studied changing recreational use of public green space in Amsterdam during the early 1990s, cites the German sociologist Bahrdt when stressing the importance of the Market as a symbolic meeting place, a well-functioning public space. The Market, or the 'marketplace', is so important that it defines the city as a sociological and economic phenomenon, in the ever-present tension between private and public spheres that a city encompasses. It requires urban behaviour; contacts are mostly quick, superficial and clear, and a certain level of anonymity remains intact. Public spaces of cities are important for marketplaces. Everyone can use them and the entire community owns these spaces. They play a crucial role in bringing people together, for example, for certain events, and stimulate face-to-face contacts between people that would otherwise hardly ever meet (cf. Van der Plas, 1991). As a public space, the marketplace has also been the place where revolutions were born, where citizens started to resist to the aristocracy and other rulers. It has thus played an important part on the democratisation of society (e.g. Clark & Jauhiainen, 2006).

Green spaces such as city parks offer the characteristics of a marketplace. While their setting maybe more 'natural', sedate and relaxed, people still often visit parks and other green spaces to meet, or at least to be with, and see other people. Gadet (1992) describes the importance of Amsterdam's green spaces as meeting places for the increasing number of singles in the city. Dings and Munk (2006) state that (Dutch) city parks are still very popular, even though they were often designed during the 19th century. The authors compare city parks with theatres, where the visitors are actors. Parks are a democratic idyll, offering a place to everyone, including otherwise marginalised groups. The best spots in public spaces such as city parks seem to be those that are not dominated by a certain social group (Van der Plas, 1991). On one hand, they are a very 'anti-urban' space that stands in stark contrast to the city. On the other hand, they are condition for urban life. This tension is acknowledged by Reeder (2006b) in the case of London. He describes how the city's parks and commons were an 'other space', "a refuge from the crowded streets of the city, promoted as 'green islands', even 'fairylands', by the London City Council. But they were also part of the multiple encounters and rhythms of city life as people from different age groups took them over for different kinds of activities." (Reeder, 2006b, p. 51).

The previous section has illustrated the mass recreational use of city forests. Like other urban green spaces, these forests have also served as social stages (e.g. Slabbers et al., 1993), but in their own, unique way (cf. Boutefeu, 2007). Even more so than inner-city parks, their charm lies in a combination of 'being away' from it all and ever-present links to city life. This important dual role of city forests is also described in Chapter 6 'The Great Escape'.

Jægersborg Dyrehave north of Copenhagen, has served as a 'social stage' for centuries. Møller (1990) describes how the various classes of the citizenry of Copenhagen went on a 'forest tour' to this forest. Preparations were intensive and while the more well-to-do had their own horse-and-wagon transportation, the poorer residents sometimes had to walk up to 15 km to get to the forest. Still, once in the forest, the different classes mixed, assembling around attractions such as a natural spring, and enjoying the entertainment provided at the 'Bakken' amusement park. This 'mixing' helped strengthen social cohesion. City forests thus became integral parts of urban life and culture, with urban life being 'transplanted' to the forest during weekends. City forests were like public parks in terms of their use, although they were mostly used during weekends and holidays. However, they also represented the more 'wild' and remote side of towns, offering people the opportunity to 'escape' from social control, as Chapter 6 'The Great Escape' describes.

Often city forests have primarily acted as a 'setting' or stage rather than an attraction in themselves. Still, even in the case of such a forest such as Jægersborg Dyrehave, where the Bakken amusement park is the main attraction, many people do come for the forest itself. Large groups of visitors assemble under the majestic oak and beech trees on sunny days, engaging in a typical bout of Danish 'hygge' (cosy socialising). In this way, people get together, tightening their social bonds, which helps to build community identity.

Social Forests and Place Making

As also mentioned by Clark and Jauhiainen (2006), green space is a vital, dynamic part of urban space. Within green space, city forests play an important social role, as they offer a range of leisure opportunities to citizens, and provide opportunities for people to meet and get together in a pleasant, semi-natural setting. When city forests act as social stages, they become sites of public cultural, social gatherings and informal get-togethers, assembling many kinds of people (Van Herzele, 2005). As with other types of green public space, the openness of city forests is vital for the emergence of a cultural identification and social attachment at the local level (Clark & Jauhiainen, 2006). By using city forests, city dwellers help shape their local environment, as urban space does not just 'exist': it is produced, reproduced and shaped by people's actions. These actions always take place in some context that is specified in space and time (Clark & Jauhiainen, 2006). Struggles over 'place making' are often mediated through culture and depend on class, gender, ethnical background, and so forth (Lepofsky & Fraser, 2003). Groups of people with similar characteristics will mostly find it easier to jointly define a place. City forests, however, attract a wide range of different people, which means that place making involving these forests requires crossing 'boundaries' of culture, gender, ethnical background, and so forth.

The process of 'place making' by people is also discussed by Lipsanen (2006), who states that not every location is a place in the full meaning of the concept. Place requires human interaction, and the more 'deep' that interaction is, the more 'placey' a place is to an individual. In his own study of the Uutela nature area near Helsinki, Lipsanen interviewed 26 visitors and found evidence of long, intensive relationships between the area and its visitors. The Uutela area had become an important part of people's identity.

Cheng et al. (2003) offer examples of 'place-based cooperation', for example by so-called 'Friends' groups that focus on a particular forest or nature area. Collaboration is generally composed of individuals who, despite their diverse backgrounds and frequently opposing perspectives on natural resource management, work together to define and address common resource management issues bounded by a geographic place.

Place making by groups of people leads to a socialisation of a certain area, meaning that groups lay claim to their places on a variety of levels, sometimes to the exclusion of other city dwellers. The case of the Arnos Vale Cemetery in Bristol, England, shows how local residents turned the area into their place, their 'territory', and started monitoring and surveillance activities (Jones & Cloke, 2002; see also Chapter 12 'A Forest of Conflict').

The (potential) role of city forests in helping to tighten community bonds and enhance place making is without a doubt a key element of city forestry, and of the contributions it can make to better cities. Communities can be described as self-defined, formal and informal groups with shared values, knowledge and interests (based on Jeanrenaud, 2001). As stated by Ponting (1991) and others, the flood of people into cities often destroyed many community bonds without creating new ones. In many urban areas, groups have become marginalised, feeling left out. Individualisation has progressed.

Olwig (1996, p. 408) links back to the concept of the commons: "We cannot improve the nature of our environment without improving the nature of our communities." As mentioned by Cronon (1996a), the problem with many new communities is that nothing seems to become really public; the primary focus is on the private domain. As a result, there is no sense that one is a member of a community. Local environment and local community go hand in hand. Groenewegen et al. (2006) acknowledge this, by stating that attractive green areas in the neighbourhood may serve as a focal point for positive informal social interaction, strengthening social ties and thereby social cohesion. Social cohesion by itself is thought to have a positive effect on well-being and feelings of safety.

Place making involving city forests should be considered in the wider context of urban 'green' planning. As urban green spaces are linked with a wide range of aspects of urban life, as described above, many different players have a shared interest (Van Herzele et al., 2005). During recent years, a call for more public involvement in urban environmental and 'green' issues has arisen. Good 'green' governance, it is argued, has to be socially-inclusive, involving those stakeholders that are affected by green decision-making, such as the planning and management of their city forests (e.g. Van Herzele et al., 2005; Janse & Konijnendijk, 2007).

The European 'NeighbourWoods'-project studied different ways of involving local stakeholders in urban woodland planning and management in case study areas across Europe. In the Bulgarian town of Stara Zagora, for example, local residents were involved in the joint development of a new management plan for the wooded Ayazmo Park, while a Swedish case study looked at letting young children participate in the management of their local wood. The project team found that a set of tools comprising a step-wise process from informing the public in an attractive way, collecting information on public opinion, towards fully participatory approaches such as direct involvement in decision-making is most likely to ensure socially-inclusive planning in city forestry (Janse & Konijnendijk, 2007). It was also noted that public willingness to participate depends on factors such as existing controversy, emotions attached to the forest and perceived dangers, e.g. in terms of threats to the status quo. Chapter 12 'A Forest of Conflict' will explore in further detail how conflicts over city forests can lead to mobilisation of (parts of) local communities.

Building Community Identity

The previous section has shown how place making, for example involving city forests, is an important part of strengthening links between people and their local environment. An interesting approach to socially-inclusive, 'green' place-making is proposed by British non-governmental organisation Common Ground (2007). This organisation aims to link "nature with culture, focussing upon the positive investment people can make in their own localities, championing popular democratic involvement, and by inspiring celebration as a starting point for action to improve the quality of our everyday places". Projects include Apple Day, Community Orchards and Tree Dressing Days, as well as a campaign to promote local distinctiveness.

Such activities build on trees as central elements of place and even as defining elements (Treib, 2002). Jones and Cloke (2002) acknowledge how plants can be powerful emblems of place and identity, not just of nations, but also of cities, villages and neighbourhoods (see also Chapter 5 'The Forest of Power'). An example of a defining tree space is New York's Central Park, which figures strongly in creating the cognitive image of the city, and the definition of its neighbourhoods (Treib, 2002). But even a single tree can provide identity, as Simon (2001, 2005) shows in the case of the Robinson Gold tree, a hybrid cypress first identified at Belfast's Belvoir Park Forest which has become a strong local and community symbol (see Chapter 10 'The Forest of Learning').

City forests have also played an important role in strengthening social cohesion, place making and creating identity through their 'memorial' roles. New city forests are established, for example, to commemorate a certain event or to start a campaign. Trees are also planted for births and deaths, and so forth, as described in Chapter 2 'The Spiritual Forest'. The children who died during the siege of Sarajevo during the war in Bosnia were remembered by the planting of trees, both in Sarajevo and in the United States (Remembering Sarajevo's children, 1997). In Belgium, a so-called 'White Children's Forest' (Witte Kinderenbos) was planted in the aftermath of a large pedophile scandal resulting in the kidnapping and murder of several young girls. The Forest symbolised public outcry and demand for radical changes in the legal and political system (e.g. Ledene, 2007).

Wars and special historical events are remembered through tree planting in both parks and forests. During the Soviet era, so-called Victory Parks were created to commemorate the Soviet Union's victory in World War II. Green space was created where the idea of city improvement was coupled with an ideological imperative of national building and promoting collectivism. However, parks and other green areas were often established too rapidly and not much thought was given to subsequent maintenance. Thus, the Soviet practices of green space construction gradually became the symbol of the unfulfilled promises, abortive plans and failed ideas of the new state (Kitaev, 2006).

Milligan and Bingley (2007) mention how nation states have drawn heavily on tree imagery in attempts to forge a national identity, for example by relating back to national folklore and myth (cf. Jones & Cloke, 2002). In Germany, forests were strongly associated with national character and culture during the 19th century, as described by Lehmann (1999) and Schama (1995). People like Wilhelm Heinrich Riehl, Germany "sociologist of field and forest" (Lehmann, 1999, p. 112), stressed that Germany's forests should be seen as essential for the country and had to be preserved. Forests were seen as the origin or heartland of German culture, given that the primeval forest had played an important role as the site of tribal self-assertion against the Roman Empire. This had been a battle of wood against stone and law, of nature against city culture. Forests were, Riehl said, what made Germany German. They could turn the 'Gesellschaft' (society) back into a 'Gemeinschaft' (community) and fight the negative trends of contemporary capitalism. This cultural meaning of the German forest featured prominently in (Romantic) art and writings (e.g. Harrison, 1992).

Forests were also used as a way to compare cultures and as a way to illustrate Germany's superiority. England and France had overexploited their forests and were thus regarded as 'lesser' countries. Russia still had extensive forests and was attributed great potential. The 'German oak' was a central point in this and oaks were gathering points in many villages and towns. Later, so-called 'Hitler Trees' were planted in honour of the Nazi leader.

In England, the 'Greenwood' played an important role in national culture and in the shaping of the power relationships, for example, between the King and his people. The forests were a refuge from state tyranny for 'rightful outlaws' such as the mythical Robin Hood, who fought against abusive rulers. On the other hand, however, the English Kings had close 'forest links' of their own, for example with the oak (e.g. Schama, 1995; Jones & Cloke, 2002).

In the United States, most of the desire to protect prominent examples of natural landscape beauty was based on a sense that the natural landscape of America was its most distinctive national heritage (Runte, 1987; Lawrence, 2006). The young country had no Roman ruins or medieval cathedrals to look to as cultural legacy. Instead, natural landscapes and their interaction with Americans acted as validation of American culture. This was reflected not only in the creation of national parks such as Yosemite, but also in parks and woods in and near cities: "the preservation of natural areas near cities and the creation of large city parks were also based in part on this sense of American exceptionalism, of a morally pure culture arising from the wilderness of North America." (Lawrence, 2006, pp. 236–237).

Nolin (2006) shows how forested landscapes played an important role in building national identity and social cohesion in Sweden at the turn of the century. City forests such as Djurgården represented the typical, idealised Swedish landscape even in the heart of large cities. Native species were favoured and nature was seen as a 'formative' force. Urban parks, as well as sports grounds, were supposed to have a morally improving influence on the working classes and to enhance physical and mental health. In early 20th century Finland, architects, planners and city officials shared a desire to create prestigious areas for the newly independent nation's capital city, Helsinki. Parks and woods helped to raise the city's prestige, and they were used for 'typical' Finnish sports and leisure activities. The city's green structure was planned and developed during a period of growing national pride, such as the run-up for the 1940 Olympics (which were cancelled because of World War II) (Lento, 2006). Tree planting, the Olympics and strengthening local identity still go hand in hand, as the example of London, host of the 2012 Olympics, shows. Extensive tree planting in East London will help with regeneration of the area, as well as leave a permanent green legacy (Nail, 2008). Another city forest landscape of importance for national identity is the Vrelo Bosne area near Sarajevo, the capital of Bosnia-Herzegovina. A wooded park, popular for recreation, has been laid out at the source of the Bosne River, from which Bosnia derives its name (Avdibegović, 2003; Fig. 11.3).

Place making and (re-)confirming community identity are also a major issue in the green belt around Brussels, Belgium. The green belt, with forests such as the

Fig. 11.3 The wooded landscape of Vrelo Bosne, the source of the Bosne river near Sarajevo, is a landscape of importance for local and national identity (Photo by the author)

Zoniënwoud, hosts an annual walking and cycling event called 'The Gordel' ('The Belt'). The event attracts up to 100,000 people. Political motivation lay behind starting this event, as it was seen as a protest against the (originally Dutch-speaking) municipalities around Brussels becoming more 'French-speaking'. It was originally hoped that the event would raise political attention and strengthen the Flemish community around Brussels. Gradually, the political dimension of the event became less pronounced (Kellner, 2000), although tensions arose again in the aftermath of the difficult Belgian elections of 2007. Also near Brussels, trees have been used in an effort to bring the Flemish and French speaking communities of the area closer together. A so-called 'Feest der Bomen' (Festival of the Trees) was organised in the Zoniënwoud during May 2007. The organising nature groups sought to involving people from both sides of the language boundary, united by their joint interest in forests and nature (Joly, 2007).

Not only national cultures and identities have been strengthened with the help of (city) forests. Cities have also turned to forests and green spaces in order to build local identity and social cohesion. In the case of 'New Towns' in Britain and elsewhere, for example, it was hoped that an extensive green structure and good local parks and woods would help build communities and make people proud of their town (e.g. Simson, 1997). Successes were achieved, although a study by Jorgensen et al. (2006) in the Birchwood district of Warrington New Town, Britain showed that, even though local woodland did feature prominently among residents' favourite places, the resource was not strongly identified with Birchwood as a place. Factors that were seen as being more important in this regard included the behaviour of local individuals, community groups and institutions.

In conclusion, the role of city forests as Social Forests is among its most crucial contributions, also today. In an urbanising and globalising society, the creation and development of local attachment, community and identity are important challenges. People need places and city forests can help in local place-making. Several examples of place making in city forestry have been described above.

Successful, joint place-making requires, first of all, that people have access to city forests and other green spaces (Travlou & Thompson, 2007; Worpole & Knox, 2007). This strengthens the use of city forests by all parts of urban society, thereby enhancing opportunities for social interaction, better social cohesion, and ultimately, joint place-making and the shaping of community identity. Accessibility of city forests relates first of all to the question whether the forest is open for public use. Many city forests are open to the public. In The Netherlands, for example, only 5% of the forest and nature areas owned by municipalities are closed off (Veer et al., 2006). However, not all countries and cities have similar levels of access. In Belgium, for example, many forests near cities are privately owned and have traditionally had limited public access (Gijsel, 2006). Access also relates to physical access, which is about physical barriers, mobility, availability of public transport to the city forests, and so forth. Finally, accessibility also has a social dimension, which has to do with people's perceptions and whether they feel spaces are accessible to them, and are not deterred by low levels of attraction, fears of safety, dominant use by another group, or simply lack of recognition of what is there (The SAUL Partnership, 2005).

A second key element of place making is the involvement of local residents in the planning and management of their city forests. Public involvement can help, for example, in developing mutual understanding between decision-makers, city foresters and residents. This type of understanding is important, for example, for avoiding or managing the conflicts over city forests that are the topic of the next chapter.

Chapter 12
A Forest of Conflict

City forests, as shown in the previous chapters, fulfil many social and cultural roles, apart from the economic benefits that they have offered over time. Chapter 5 'The Forest of Power', however, already showed that is it impossible to meet all demands for forest goods and services with the often rather small city forest resources available. Moreover, people have different opinions about what city forests should look like and how they should be designed and managed. Given the large interests in city forests, for example because of their role in place making, and the many societal demands, it is obvious that conflicts of interests are common. This chapter looks at city forests as Forests of Conflict.

Folger et al. (1997, cf. Walker & Daniels, 1997) define conflict as the interaction of interdependent people who perceive incompatible goals and interference from each other in achieving these goals. Conflicts can be substantive or content-based. They can also be procedural, relating to 'rules' and procedures, or personal, i.e. between actors, regarding issues of power, authority, control, identity and the like (Walker & Daniels, 1997). Public parks are typical sites of contestation and resistance (cf. Jones & Cloke, 2002). 'Tree places' are characterised by Jones and Cloke as landscapes whose material form actively incorporates both the struggles which have occurred over it and the very nature of these struggles. City forest history is also full of social conflicts, which is not strange given the (urbanisation) pressures placed upon these forests over time and the many different demands of city dwellers.

Chapter 5 'The Forest of Power' considered how different societal classes and groups have attempted to appropriate city forests for their own benefit. Obviously this leads to conflicts, as was also shown in that chapter. Rather than integrating modern-day city forest conflicts in Chapter 5, they have been given a chapter of their own. An important reason for this is that city forest conflicts can hamper planning and management, and act as important obstacles for realising the Forests for the Future that are described in the final chapter of this book.

Among the historical conflicts over city forests and city forest use described in Chapter 5 are those between aristocracy and the lower classes of society, with the conflicts over the Haagse Bos and the London commons as eminent examples. Conflicts have often been over gaining access to city forest and over user rights. This chapter looks beyond these aspects, focusing on city forest conflicts that very much determine current planning and management. With the democratisation of

C.C. Konijnendijk, *The Forest and the City: The Cultural Landscape of Urban Woodland*

society, more people have been demanding influence in decisions over city forests and their use. As a result, the number of conflicts has increased further. In the next sections, examples of city forest conflicts will be presented according to some main groupings: urban development conflicts, conflicts over city forest management, recreation conflicts, as well as other conflicts. The final section attempts to go 'deeper' into city forest conflicts, relating them to the often close links between people and city forests as places. The section shows that comprehensive conflict management strategies can help turn conflicts into a positive force.

Urban Development Conflicts

As described in the first chapters of this book, forests often fall victim when cities develop, even though great value is often placed on city forests. Yet, when new roads need to be constructed, or new housing has to be developed, the eyes of politicians still frequently turn to city forests. This was already the case when the population of the Dutch city of Arnhem protested against the plans of the city to establish expensive housing in parts of the wooded Sonsbeek park (Kuyk, 1914). Elsewhere, the so-called Djurgården Parliamentary Committee of 1905 considered Norra Djurgården in Stockholm very suitable for establishing new dwellings. Botanist and politician Karl Stärback led a countermovement, offering a petition to Parliament to preserve the area. The area's importance for the citizens' recreation was emphasised and reference was even made to the need for the Swedish capital to have a well-preserved, original landscape of a typically Swedish character (Nolin, 2006).

More recently, there has been debate in Helsinki, Finland, about where to build new houses. Strategies for keeping the city open have been succeeded by infill strategies (Niemi, 2006). Citizens often protested when building plans and activities threatened green areas, as in the case of Vuosaari. The Las Kabaty forest of Warsaw has also felt the pressure of development, as a new city district for 150,000 people became its neighbour during the 1970s. The city's Nature Conservation Bureau managed to have the forest placed under the nature protection law, after a long and difficult process. This meant that the forest obtained a double protected status, under both forest and nature conservation law. Still, urban development has been affecting the forest, as some houses have been built or expanded into the legally defined buffer zone surrounding the forest (Konijnendijk, 1999). In the forests near St. Petersburg, Russia, development conflicts regularly arise when trees are being cut for establishing dachas (summer houses) and permanent housing for the new rich. In cases like that of the protected area of Lindulovskaja Roshcha, political corruption led to the bypassing of nature protection regulations (Konijnendijk, 1999).

Similar cases of city forest versus urban development are also known from other parts of the world. On a more historical note, forests on the hills near the new city of Portland, Oregon, were destined to be cut and developed. During the second half of the 19th century, however, the clergyman Thomas Lamb Elict, serving as on the city's first Municipal Parks Commissioners, started to work for conserving at least part of these forests. Continuous political debates and actions such as tree planting activities, as well as support from part of the business community, led to the establishment of Portland's Forest Park, currently one of the largest city parks in the world (Houle, 1987).

The development of new residential areas has not been the only type of urban development pressure threatening city forests. Infrastructure development has been another major driving force. In The Netherlands, the forest of Amelisweerd near Utrecht was partly sacrificed to the construction of a new highway, but the protests against it in the 1970s and early 1980s were fierce and delayed the construction with several years. The case of Amelisweerd would later be regarded as the start of a strong nature conservation and public action lobby in The Netherlands (Grimbergen et al., 1983). The British government came up with a plan for an orbital highway (the M25) which had to go through the Epping Forest. But the City of London did not allow this to happen (Slabbers et al., 1993). However, during the mid-1990s the Newbury Forest could not be protected from the governmental plans. A part of the forest was cut during 1995 and 1996 for the construction of a new highway near London. Emotional and fierce protests of action groups – who established entire villages in the forest and had people climbing in trees – could not prevent this from happening (Konijnendijk, 1996).

A very recent case illustrates that conflicts involving city forests and infrastructure development can be complex. Inhabitants of the Polish town of Augustow had complained for years about the heavy traffic, including up to 4,000 trucks per day, passing through the centre of their town. In response, the Polish government decided to establish a new ring road. It was also agreed that part of the new road would cross the nearby Rospuda Valley, a natural area comprising woodland and river landscapes. Nature conservation groups objected against these plans and occupied some forests in the valley. The European Union also got involved, as the valley is considered an area of nature beauty of international standing. A European Court ruling called for road building, which had already commenced, to be stopped. In this particular case, the inhabitants of the city involved were taking the side of the road developers, opposing the opinions of the nature conservation movement (Aagaard, 2007; Hunin, 2007).

Another example of conflicts between economic development and city forests is the case of National Park Donau-Auen situated in and near Vienna. Actions by nature conservation groups and a general public outcry prevented a hydroelectric plant from being built in the Park (A. Ottitsch, 1997, personal communication). There was less success for a Dutch environmental NGO attempting to block the expansion of Schiphol airport near Amsterdam. It did so by planting a 'forest' – the so-called 'Bulderbos' – on the very site where the future runway expansion would be situated. The first trees were planted in 1994. When the courts ruled that the forest was in fact illegal, its trees were transplanted – or, in the words of the NGO, 'sent into exile' – to a city forest near Almere (Milieudefensie, 2007).

That brings us to a recent city forest conflict case that in many ways is characteristic for the struggles between urban development and city forest conservation, namely that of the Lappersfortbos, a 30 ha woodland near the centre of Bruges, Belgium. This woodland was once a private park owned by the director of an arms producing company. The grounds, including adjoining warehouses, were bought by a multinational company, with the aim to partly develop it into an area of buildings and infrastructure. For this, a land use plan was adapted which designated two-thirds of the former industrial area as development area. One third of the former industrial area was set aside as a park area.

After approval of the plan by the city council of Bruges, and after the usual public consultation processes, preparatory works which involved extensive cutting of trees were about to start. However, a group of mostly young people from Belgium and others parts of Europe who objected to the cutting occupied the forest. Their tree houses remained in the area for more than a year, even though a court ruled in favour of removing the protestors. A long stand-off and media campaign started and the Lappersfort-case developed into the 'darling' of the Flemish green movement (e.g. Lappersfortbos vandaag of morgen, 2002; Groene Gordel Front, 2006).

During the time of their actions, a sort of subculture developed amongst the protestors, very similar that what had happened during the previously mentioned forest occupations at Amelisweerd and Newbury. The protestors told the press about their 'transformation' from city people to real forest dwellers. The protestor community became well organised, with a clear division of responsibilities and tasks. Some people who allegedly 'came for the wrong reasons' were evicted. Newcomers were told that the Lappersfort-camp was far from a holiday camp. Nobody lived in the forest for a full year, as the general feeling was that one needed to 'disconnect' from the forest every once in a while. Also, the risks of having difficulties in returning to society afterwards were considered. Huts became more professional over time, as did facilities. A first 'Lappersfort-baby' was soon on its way (De eerste echte, 2002). Visitors came to admire the 'green occupiers', who were ascribed with a 'Robin Hood-like charm' (Prijs Lappersfortbos nog, 2002).

Finally, the protestor camp and its last 35 occupants were removed by 120 police officers. The site was prepared for the demolishing firms with their heavy equipment. The media compared the action with a real siege and the look of a war zone. The owning company went to court, with the aim to charge the protestors with the cost of the removal operation. The court granted this claim. Immediately after destruction of the camp, a private security firm stood guard against new occupations (Politie ontruimt Lappersfortbos, 2002).

Finally the case seemed to come to some sort of happy ending in 2003, when the Flemish Minister of Environment reached an agreement with the owner about buying the Lappersfortbos. However, the company only agreed to sell at a price appropriate for the land designation (i.e. as industrial area), as it wanted to be compensated for the losses incurred for not being able to develop a business area (Dua heeft voorakkoord, 2003). At the time of writing, a final settlement had not yet been achieved, primarily due to the resistance by the Flemish Minister of Finance against the exorbitant price for buying a forest. The protestors thus remained alert and active as the threat of the deforestation is still there. In 2006, a book with poems about the forest and the times of the occupation was released by the 'Lappersfort Poets Society'. An online 'Lappersfort museum' (Groene Gordel Front, 2006), an archive with stories, diaries and poems, was maintained. Some of the protestors became active in an international network that protested against deforestation in other parts of Belgium and in The Netherlands (R. De Vreese, 2007, personal communication).

Conflicts over Forest Management

New roads, buildings, and other threats to urban forests often bring inhabitants, environmental groups and city forest managers to the same side of the conflict. City forest managers may also find themselves opposing the public, however. Ronge (1998) mentions how during the last few decades, ecological commitment and knowledge have developed considerably amongst lay people. As a result, the professional legitimacy of foresters is often denied. In other words: foresters are not necessarily seen as the most qualified persons to decide on what will happen with the (city) forest. As described in Chapter 8 'The Wild Side of Town', the first citizens who expressed environmental awareness came from the upper layers of society and from academia. This was the case, for example, when the Swedish Society for Nature Conservation was established in 1909 (Nolin, 2006). Founding members came primarily from Stockholm and surroundings and where deeply concerned with the city's flora and fauna. They started to argue with municipal authorities about the design and management of Stockholm's green areas, as they wanted to see more emphasis on nature and natural conditions. Another case from those days, which did not only involve the upper classes, was that between forester Van Schermbeek and the inhabitants of Breda, The Netherlands, over management of the Mastbos city forest. Van Schermbeek introduced northern red oak (*Quercus rubra* L.) and removed the undergrowth in a (failed) attempt to ameliorate the dehydrated soils of the Mastbos. He then ran into conflicts with members of the public, including blueberry pickers, who saw their source of additional income disappear with the changes in forest structure (Caspers, 1999).

The forest of Fontainebleau in France has been a prime example of conflicts over forest management between professional foresters and others. Several previous chapters have illustrated how a group of artists protected successfully against the cutting down of oldest part of the forest for regeneration with pine by Forest Administration during the early 19th century. But also in modern times, fierce conflicts have arisen over Fontainebleau and its management. During the 1990s, nature conservationists claimed that the French state forest service (the Office National des Forêts) still placed too much emphasis on timber production, while not giving sufficient attention to nature conservation aspects. Subsequent actions of a radical part of the conservationists, an environmental group labeled as 'eco-terrorists' by some and as modern Robin Hoods by others, included destruction of machinery, material and plantations, and the painting of action slogans on the houses of policy makers. Several of the protestors found themselves in prison after this (Ruffier-Reynie, 1992; Brésard, 1995; Trébucq, 1995).

Elsewhere, a rather similar conflict occurred. The Forest Service of Oslo regularly became the main target of protests as leisure seekers and nature conservationists who considered timber operations – including the construction of new logging roads and large-scale clear-cutting – to be destroying the Oslomarka forest. Public criticism against forestry especially grew from the middle of the 1950s, when logging road construction as well as regeneration increased. At first, the forestry profession strongly rejected the demands as expressed by the public, believing that criticism originated from only a small group of people (Hellström & Reunala, 1995). During the Nature Conservation Year of 1970, Oslomarka was made into a major political issue. In 1972, environmentalists tried to prevent the construction of a logging road by placing themselves in front of the machines. Finally, the parties came closer to each other and forestry policies and practices were adapted according to recreation and nature conservation objectives (Opheim, 1984; Hellström & Reunala, 1995). In a way, the 'Oslomarka-affair' set an example for forestry in entire Norway. The country's Forest Act was renewed in 1976 so that issues of multiple use could be incorporated. This shows that city forest conflicts can be a positive driving force as well, a topic which will be given more attention later.

Differing nature views and perhaps lack of public awareness about ecology and forestry are an important source of conflicts over city forest management. This is illustrated by the case of Germany during the 1980s, when the forest dieback ('Waldsterben') debate was at its peak. There were continuous discussions in the media and calls were even made for abandoning the tradition of having Christmas trees. Urban dwellers were making emotional, 'final' visits to their local, 'dying' forests (Lehmann, 1999). Similar reactions were much fewer, however, amongst residents of smaller towns in the countryside. These people were in more regular contact with nature and failed to see dramatic impacts on the forest, which reduced their concerns and willingness to protest. This case relates to the conflicting urban and rural nature views mentioned by Fairbrother (1972). These conflicting views also manifest themselves when urban dwellers visit the surrounding countryside and disagree with rural practices, such as those of rearing farm animals.

Elsewhere, many other conflicts over city forest management have been described, such as those over the Keskuspuisto woodland park in Helsinki, where nature conservationists wanted to have part of the forest set aside for nature protection purposes (Konijnendijk, 1999). In some cases, conflicts are primarily between forest managers and environmental groups, who mostly want less management and more natural processes. In other cases, local inhabitants are leading the protests against forest management, for example when they feel that they have not been properly informed about thinning of harvesting operations, forest regeneration and the like. Local inhabitants are also the main recreational users of city forests. This leads to their involvement in yet another type of city forest conflicts.

Recreation Conflicts

Even more typical 'day-to-day' conflicts in city forests are those involving different types of recreational use (Fig. 12.1). In a series of interviews with city forest managers in 16 European cities, for example, mountain biking (Fig. 12.2) was mentioned by almost all interviewees as a problematic activity that hampered walking and other recreational uses (Konijnendijk, 1999). A user study in the city forests of Vienna indicated that visitors saw dogs off their leash as the most 'annoying factor' in the forest, besides litter and vandalism (Arnberger et al., 2005). Regular

Fig. 12.1 City forest managers have restricted some forest uses, for example to reduce conflicts between different types of recreation, as well as between nature conservation and recreation. In the Zoniënwoud near Brussels. for example, signs indicate that motorised traffic and picking mushrooms are not allowed (Photo by the author)

visitors often object to large-scale events in 'their' forest. This was the case of the Sonsbeek woodland park in Arnhem, where the 'Friends of Sonsbeek' group successfully protested against the hosting of the 1995 World Liberty rock concert in the park (Sonsbeek krijgt ruime, 1996; Van den Ham, 1997). In general, large-scale events such as marathons are not very popular with many regular visitors.

Particular, traditional forms of recreational or tourism use of city forests can also lead to protests when times change. According to a long tradition, inhabitants of communist East Berlin had been able to set up large tents in the forests. People lived in these tents for weeks and often even months during summer. After the German reunification, this caused a problem, as this type of activity was not

Fig. 12.2 Mountain biking has been the cause of recreation conflicts in many city forests, for example in The Netherlands (Photo by the author)

allowed according to the law of Federal Germany. The Berlin Forest Service was also concerned that the large amount of tents in the forest made other forms of recreational use impossible in some forest areas. As the tent associations were too poor to buy any land, and as the tent (Zelten) culture was so deeply rooted, the policy of the Forest Service was to gradually end this practice, but not by enforcing tent owners to leave (Konijnendijk, 1999).

In countries such as The Netherlands, private forest owners receive subventions for opening their forests to the public. However, forest owners are not always happy with recreational use, as it reduces their privacy and can cause damages to the forest and its infrastructure (Konijnendijk, 1999; Fig. 12.3). A recent case from Denmark, although not strictly a city forestry case, illustrates that emotions can run high. A Baron who owns a wooded estate on the Danish island of Gavnø recently blocked all access roads to the estate, as he wanted to keep leisure seekers out. However, he was not allowed to restrict public access according to the Danish forest law. The Baron even posted a bull behind a main fence to scare away visitors. He considered taking his case to the European Court of Human Rights, as he felt that the nobility should be allowed to enforce their traditional private property rights (Stisen, 2003). Power struggles and conflicts between forest owners and users, one of the central themes of city forest history, thus persist in modern times.

Fig. 12.3 Private land owners are not always happy with the recreational use of their forests and lands, as this sign in Ireland shows (Photo by the author)

Other Conflicts

Contemporary city forest conflicts often involve threats from urban development, differences of opinion over forest management and different recreational uses, as described above. But other types of conflicts also occur. The use of forest products leads to conflicts, as in the case of poaching and the illegal removal of timber and protected species such as mushrooms. A rather 'new' conflict is that between agriculture and forestry. In fact the conflict as such is not new, of course, as forests over time have fallen victim to agriculture. Today, however, the trend often is the reverse, as agricultural land near cities is planted with forests. The Danish Vestskoven ('West Forest') west of Copenhagen, for example, was established during the 1960s and 1970s on agricultural lands. The planning and establishment of the forest led to protests by farmers and a few of them have held on to their land, now situated in the midst of the Vestskoven. Efforts to establish city forests near forest-poor Belgian cities such as Ghent, Kortrijk and Roeselaere have also been hampered by farmers not willing to sell their lands (e.g. Vitse, 2001; Van Herzele, 2005).

Another group of city forest conflicts comprises those between nature conservation, recreation and other interests. Nature conservation organisations would sometimes like to see parts of city forests with high natural values excluded from (at least intensive) recreation. The private Iskælderskoven, a forest in the Danish municipality of Fugleberg, is a rare breeding site for the golden eagle (*Aquila chrysaetos* L.). The local nature conservation association succeeded in getting authorities to close off the forest for recreation for three years. Recreationists, represented by the Danish Outdoor Council, objected, as they feared that this would set a precedent for similar cases elsewhere (Den her skov, 2006). Disagreement over large predators such as wolves and lynx, and their reintroduction even to areas close to cities, has also led to conflicts in countries such as Germany, Norway and Sweden (e.g. Hagvaag, 2001).

The efforts of different actors to claim ownership over city forests has been described earlier in this chapter, as well as in Chapter 5 'The Forest of Power'. Over time, city authorities have increased their influence over nearby forests. In the former communist countries of Eastern Europe, municipal property was transferred to the state. When the Iron Curtain fell during the late 1980s and early 1990s, a long and troublesome process of reconstitution of formerly municipal and private forests started (e.g. FAO, 1997). Bisenieks (2001) mentions how the Latvian city of Riga gradually reclaimed its more than 70,000 ha of forests and agricultural land – part of which can be traced back to city ownership in the year 1225 – after the country had regained its independence.

Managing the Forest of Conflict

Conflicts thus are not only over competing uses, but rather over (differing) meanings and bonds between people and landscape. Many city forest conflicts can be better understood from a 'place perspective' (Cheng et al., 2003), as they concern the place identity of individuals as well as groups and communities. Identity, of course, has a powerful influence over people's behaviour. In the case of groups and territoriality, people feel that outsiders impose on 'their' territory, threatening their very identity as a group. Groups who 'claim' a city forest as their place, their territory, have certain expectations of appropriate behaviour assigned to this place. However, in the case of city forests, which are mostly multipurpose public forest land, expectations of appropriate behaviour are not well defined and well accepted. This means that conflict is a pervasive feature of city forest planning and management decision making (based on Cheng et al., 2003).

Conflicts such as those described in this chapter can stand in the way of successful city forestry, challenging forest planning and management. On the other hand, conflicts are a sign of people's commitment to 'their' city forests. The example of the Lappersfortbos, where protests were very emotional, shows clearly how strong the ties between city forest and people can be.

Conflict management is about making use of the positive dimensions of conflicts and reducing the negative impacts. Walker and Daniels (1997) stress the importance of referring to conflict 'management', rather than conflict 'resolution'. Conflict management, they say, is aimed at situation improvement. It starts from the premise that desirable and feasible changes can be made in any problematic situation.

Conflict management considers the 'deeper' roots to conflicts, looking at substance of conflicts, as well as procedural and relational aspects involved. People often feel by-passed, for example, when decisions about city forests are taken without their involvement or even their knowledge (e.g. Konijnendijk, 1999). Thus conflicts with decision makers and forest managers can arise, even when there is agreement about the proposed planning and/or management actions. Relational aspects of conflicts have to do with the relations between individuals. If people do not like or trust one another, conflicts are more likely to occur.

The close links between people and their places also need to be taken into account when conflict management strategies are developed. This requires a sound understanding of how different groups and individuals in a local society relate to the local city forest. One of the worst things to do is to ignore the interests of those who have deep relationships with their forest, considering them as part of their place identity. With an appropriate conflict management strategy, escalation in the Forest of Conflict can be avoided. Moreover, people's commitment to the forest can be used for the better. It can help withstand political pressures to convert city forests to other land use, for example, and raise resources to support forest management.

Chapter 13
A Forest for the Future

This book has described and analysed city forests as cultural landscapes. It has shown that city forests have, over time, not only played important economic and environmental roles. These forests have particularly also had many cultural and social values. Starting from their earliest role of providing subsistence to the poor, city forests became appreciated for offering a wider range of goods and especially services to local urban communities. Over time, the use of city forests became more 'democratic', as the exclusive rights of the elite were transferred to the urban public at large. City governments enhanced their influence over city forests, as they could see the benefits these provided, initially in terms of supplying crucial timber and fuelwood and protecting drinking water resources, and later for creating attractive environments for residents and businesses as well.

This final chapter will look once again at the role of city forests in the construction of landscapes, places and spaces. It will start by analysing the challenges of the present urban era, where large parts of the cultural landscape have become urban or at least urbanised. With ongoing urbanisation, suburbanisation and urban sprawl, as well as processes such as globalisation, cities and city regions are in need of stronger identities. They have to compete for inhabitants, businesses and visitors in the global marketplace, not least by offering a high quality of life and environment.

City forests can help in creating competitive cities by being part of attractive, multifunctional urban landscapes. Although their integrity and own (cultural) identity need to be recognised, city forests must be integrated in more strategic planning and management of the cultural landscape. Several examples of how this is done today will be considered. New concepts, such as that of 'green infrastructure', can help in enhancing landscape integration and functionality.

Among the challenges of cities and urban areas today is the need to maintain and improve social cohesion. How can connections between people and their locality be promoted, at a time where urbanisation seems determined to sweep away the differences between regions, cities, towns and landscapes? How can local roots and local identity be strengthened, so that people feel at home and committed towards their local area? How can newcomers be integrated? Questions like these all relate to the role of certain landscapes as 'place'. As we have seen in Chapter 11 'The Social Forest', city forests (can) play an important role in the making – or re-making – of place.

C.C. Konijnendijk, *The Forest and the City: The Cultural Landscape of Urban Woodland*

It is not only through their place-making contributions, however, that city forests can assist in meeting some of the challenges of modern urban society. There is also a need to maintain links with nature, with 'the other', even in the most urbanised of environments. This is where the role of city forests as 'space' comes in. City forests, as we have seen in Chapter 8 'The Wild Side of Town', offer an opportunity to experience nature and wildness nearby, as a contrast to urban life. In this sense, city forests are important learning environments, not least for the younger generations, as explained in Chapter 10 'The Forest of Learning'.

By analysing the (potential) roles of city forests in multifunctional urban landscapes, in place (re-)making and in keeping nature on people's doorstep, this chapter provides some thoughts on the future of our city forests. If the offered opportunities are grabbed, true Forests for the Future can be created.

Challenges in an Urban Era

During the early 1970s, Fairbrother (1972, p. 181) characterised modern culture as being metropolitan. "Whether we like it or not, and whatever setting we live in, the big city is the unit of modern life and the basis of industrial societies." Interesting in the context of this book, Fairbrother (1972, p. 199) compared the big city, with its assembly of separate, human-scale places, with a forest: "We do not, any more than in large forests, encompass it all at once; we cannot see the woods for the trees and we therefore live in the small-scale intimacy of our own forest clearing." It is true that urbanisation has led to dramatic changes in our societies and in our landscapes, particularly in the industrialised world. From a landscape where isolated, mostly small cities were situated in vast areas of wilderness, forest and later countryside, societies have moved to a primarily urban landscape, where most of the remaining nature and countryside has come under urban influence (e.g. Fairbrother, 1972). Suburbanisation and sprawl are expanding this urban influence, eating away at the remaining open land. Large parts of our landscape have started to look very similar.

In a recent article, Allman (2007) tells the story of the American city of Orlando, Florida. He writes (p. 99): "The Orlando region has become Exhibit A for the ascendant power of our cities' exurbs: blobby coalescences of look-alike, overnight, amoeba-like concentrations of population far from city centres." Urbanisation and suburbanisation areas like Orlando, Allman states, have major problems in creating an identity of their own. From a swamp, Orlando developed into a 21st century metropolis in no time. A major factor in its development was the creation of the Disney World theme park during the 1960s. Disney World made the city known all over the world and a popular destination for tourists. More is needed, however, to give the metropolis an identity of its own among its dwellers, many of whom have come from all over the world in pursuit of better lives.

As discussed in Chapter 1, successful cities today must resonate with the ancient fundamentals of being sacred, safe and busy (Kotkin, 2005). Being sacred can be linked to a joint identity and to people feeling proud of their city. Safe cities are places where people enjoy living and working, without too many concerns about their safety, health or well-being. Busy cities, finally, are full of business, cultural and other activities. Modern times have also led to the rise of the ephemeral, the dynamic city, where entertainment and new experiences are sought. Orlando can act as a symbol of this type of city. However, successful cities cannot depend only on this. For more sustainable success, they need an engaged and committed citizenry. A set of factors is thus needed, a mix of the aesthetic and mundane, of the one-off and long-lasting.

Although the megacity still plays an important role today, the balance of power between different types of urban area is shifting. Megacities are sometimes in crisis, as these "giants are losing out to smaller, better-managed, and less socially beleaguered settlements" (Kotkin, 2005, p. 146). New technologies are diminishing the advantages of scale and strengthening the trends of decentralisation. In line with this, Foroohar (2006) speaks of the rise of the Second City. With most urban growth now occurring outside megacities, Second Cities from exurbs to regional hubs, resort towns and provincial capitals are gaining prominence. Further sprawl is expected, as high land and house prices in the urban core and traditional suburbs drive people to more distant exurbs. Second Cities require excellent transport – think public transport, low-cost airlines – and communication links in order to succeed (cf. Gehl, 2007). The 'good', cosmopolitan life, formerly a matter of the large cities, has become democratised and decentralised. Suburbanism is thus very much alive, as real growth in jobs and populations is likely to take place on the periphery. In a country such as Britain, about half of the population already lives in suburbs or 'exurbs'. Kotkin (2006) argues that it is time to make suburbanism work better, rather than to just fight it. Suburbs, he says, should be made into small cultural, social and economic centres of their own. The rise of the Second City points in this direction.

Urbanisation in a wide sense holds the risk of making "places citylike without necessarily making cities" (Gallagher, 1993, p. 19). Zijderveld (1983) mentions lack of 'urbanity' as an important problem for many cities today. Urbanity is defined as the cultural character of a city which gives inhabitants a collective identity and a feeling of joint belonging. Referring to the work of German sociologist Hans-Paul Bahrdt, Zijderveld mentions how community ('Gemeinschaft') has gradually become replaced by a society with much looser and less personal ties between its people ('Gesellschaft'). In a time of globalisation, urbanisation, commercialisation and privatisation, where social cohesion is under pressure, there is a need to regain urbanity and to recreate a strong civic society. Modern urban societies are multicultural and new residents have brought along new expectations, for example about public spaces in cities (The SAUL Partnership, 2005; Walraven, 2006).

A lack of urbanity and local identity is a problem, as our life is getting so complex, with so many different social roles, that places are needed to support rather than fragment our lives, as the more anonymous urban areas may do (Gallagher, 1993). People need a real home, a place that roots them in a hectic, rapidly changing world.

Local identity is called for, as are "places that balance the hard, standardised, and cost-efficient with the natural, personal and healthful" (Gallagher, 1993, p. 19). A consortium of city regions in North-western Europe that jointly undertook European Union-funded sustainable development projects, writes in its final report: "The forces that formed our urban landscapes in a post-industrial era, both economic and social, have also led to a loss of traditional regional identity." (The SAUL Partnership, 2005, p. 16). Loss of former industries, for example, has created a loss of the strong communities that many localities used to enjoy.

What can be done to keep or make cities, suburbs and other urban areas successful? How can they be made into places that have a high quality of life and environment, as well as a clear identity of their own? Sustainable urban development requires a balancing act between economic, ecological, social and cultural dimensions. Quality of life refers to a wide spectrum of cultural, educational, and other possibilities offered to the population. The concept of a 'good' city today is related to aspects such as sustainable economic growth, social cohesion and equity, and a balanced ecosystem (e.g. Olin, 2007). It is about creating humane, attractive, liveable and habitable cities enabling people to identify themselves with the locality where they live (European Science Foundation, 2004). This requires strategic and innovative approaches to urban development.

Developing Attractive Urban Landscapes

In its revised Lisbon Strategy (European Union, 2007), the European Union focuses on the issue of competitiveness in a globalised society. City regions, as economic powerhouses, are crucial in this respect. But they have to constantly compete for political attention, investment, residents and visitors. Some city regions are struggling with the relicts of the past, such as the abandonment, dereliction and related negative image resulting from old (heavy) industries. Some regions are expanding, while others are contracting, losing population.

The earlier mentioned SAUL Partnership (2005) calls for new urban landscapes, with a vital impact on people's quality of life in city regions. Sustainable regions are ones where people want to live, now and in the future. A joined-up approach is required to develop these sustainable city regions, bringing together different geographical and stakeholder interests. "Competitive city regions are ones that can attract and retain viable businesses and their employees by offering a good quality of life. New urban landscapes are an essential element in building Europe's future economic structures and social well-being" (The SAUL Partnership 2005, p. 1).

How can city forests contribute to these new, sustainable urban landscapes? First of all, most city forests are an important part of public open space. The SAUL Partnership stresses that decision-makers should be aware of the importance of high quality, accessible open spaces, as all citizens need access to well-managed spaces that are clean, safe and fit for purpose in the 21st century. Second, in many city regions forestry can be a major contributor to multifunctional urban landscapes.

The traditional 'Fruitful Forest' (see Chapter 4), providing timber, biomass and other products, can be combined with the provision of a wide range of social, cultural and environmental services.

Lawrence (2006) reflects upon this as he expresses some thoughts about the modern role of urban trees in general, based on his historical analysis. Trees, Lawrence writes, are still used as ornaments or to hide architecture and urban engineering. Moreover, they continue to carry with them associations of pleasure, as described in Chapters 6 'The Great Escape'. Urban trees also still help to create spaces for 'genteel behaviour' and for displays of wealth and taste. However, with changes in urban society and lifestyles, the role of trees and woods has also changed. This has to do with wealth and taste being often displayed today through material possessions. Public landscapes are under pressure with the growth of the private sphere.

However, a new role of trees and city forests in particular has emerged. They have become instruments and even symbols of the fight against dereliction and urban decay, of the development of high-quality urban landscapes (e.g. Lawrence, 2006). The cores of metropolitan regions, and in particular their central cities, have been struggling with the dereliction of buildings and infrastructure, challenged by falling tax bases to maintain what they have, let alone create new. Revitalisation efforts in many parts of Europe, for example of abandoned industrial lands, often involve generous plantings of trees. Jones and Cloke (2002, p. 57) speak of the "transformative action" of trees that is enrolled by forest and land planners to produce both material and cultural effects on local and visiting populations.

In this context, the so-called brownfield sites present promising opportunities for city forestry, as can be derived from recent European projects such as CABERNET, a network focusing on brownfield and economic regeneration (Ferber et al., 2006). Brownfield sites are defined as "sites that have been affected by the former users of the site and surrounding land; are derelict and underused; may have real or perceived contamination problems; are mainly in developed urban areas; and require intervention to bring them back to beneficial use" (Ferber et al., 2006, p. 3). These sites can stand in the way of effective urban land use and thus of sustainable city regions by, for example, having a negative impact in terms of image and attractiveness on the surrounding area and the community. However, brownfield sites also offer opportunities. Their regeneration can stimulate opportunities at numerous levels to improve urban quality of life, enhance urban competitiveness and reduce urban sprawl. In the light of current debate on shrinking cities and declining economic resources, the creation of urban woodlands may be a particularly attractive way to regenerate brownfield sites (Burkhardt et al., 2007). Brownfield sites create opportunities for new city forests in inner-city locations, where new forests are most commonly established on the urban fringe.

One example of a new city forest landscape on a former industrial site is the Parco Nord, situated 9 km from the centre of Milan, Italy. Since the 1980s, more than 600 ha of brownfield land have been converted into a landscape mosaic comprising woodland, agricultural land and other open spaces in combination with infrastructure (such as an airport). The Parco Nord is now part of the 'Metrobosco' initiative, a larger, ambitious project to afforest thousands of hectares of land in the Greater Milan agglomeration.

The project has generated important knowledge on how to establish a wooded landscape on abandoned industrial land, thus also acting as a 'Forest of Learning' (Gini & Selleri, 2007; Sanesi et al., 2007; P. Valentini, 2007, personal communication).

Although the classic dichotomy of city and landscape in urban regions no longer exists, the image of clear separation of built-up areas and nature still dominates our understanding of spatial planning (The SAUL Partnership, 2005). It is therefore important to think beyond traditional boundaries when planning, developing and managing landscapes. Fairbrother's (1972) 'green-urban landscape', combining a focus on recreation with industrial components, comes to mind,. Hiss (1990, p. 127 etc.) cites the early 20th century work of American forester and planner Benton MacKaye, who stressed the importance of planning at the level of a 'regional city'. In his book 'The New Explorer', MacKaye mentions three main elements of a regional city: the primeval (i.e. more or less 'wild' nature), the urban and the rural. These three elements need to be balanced and connected. For the rural component of the 'regional city', Birnie (2007) describes how over the past 50 years a shift has taken place from landscapes for food to landscapes for fun. The consumer agenda for rural areas – and particularly for those in and near city regions – has become concerned with issues such as conservation, amenity and recreation. This new form of cultural landscape is often owned by a conservation, amenity or recreational trust or a local community group, or indirectly supported through agri-environmental or rural stewardship payments to the land-owner.

In an article about new planning approaches for Asian mega-cities, Yokohari et al. (2000) also stress that we need to look beyond the urban-rural divide. What urban areas need, the authors argue, is a controlled mixture of urban and rural land uses, a mosaic that provides a wide range of ecological, social and economic functions. The traditional hillside coppice woods of the Japanese 'satoyama' landscape, for example, have become an important and much appreciated element of many urban agglomerations (e.g. Takeuchi et al., 2003). They still have a role as Fruitful Forests, as the coppice woodlands are a source of wood for bio-energy. The landscapes act as Wild Side of Town as well, hosting high levels of biodiversity. Moreover, the satoyama landscapes offer a Great Escape to urban dwellers (e.g. Gathright et al., 2007), as well as a spiritual and educational link to the past.

Landscapes such as that of the satoyama may help find a careful balance between urban and rural areas and characteristics. Dietvorst (1995) argues against a complete transformation of rural areas in urbanised societies to 'new' nature and forests. Rural landscapes can be of high cultural-historical importance, he writes. Our present culture is characterised by acceleration and compression of time and space, with elements such as newsflashes, flexible jobs, time-share. As a result, a longing for the past has emerged. Landscapes of a more rural character offer opportunities for recreation and tourism. Like rural agricultural areas, older city forests also represent a link to the past.

As can be derived from the above, city forests should be considered not in isolation, but as important part of multifunctional urban landscapes, representing (in the words of MacKaye) the 'primeval', the rural and – to some extent – the urban. The parts of these urban landscapes where woodland and trees determine character and functionality could be called 'urban forest landscapes'. These typically include wooded areas, but also other types of land use, as in the case of the English Community Forests, which strive for a tree cover of 30% or so, and the previously mentioned Parco Nord. Simson (2005b, also 2007a) explores how successful urban forest landscapes can be created, describing success stories of larger-scale, diverse and multifunctional landscapes in which forests and trees are only one element, albeit a very important one. One of these examples is the Emscher Landscape Park, a forested landscape in a heavily urbanised Ruhr area of Germany (Fig. 13.1). In 1956, close to 150 coal mines still employed half a million people in the area (Neiss, 2007). Many foreign immigrants came to work in the mines, as well as in the steel and chemical industry. Now, a 450 km^2 large post-industrial landscape is under development along the Emscher River. Old production areas are developed into new, natural landscapes and new 'wilderness'. The slogan used for turning a former industrial landscape into an urban forest landscape has been 'bringing the past through the present into the future'. This shows the importance of exploring and strengthening cultural links, in this case the former industrial use with a new 'green' and recreational use in which industrial heritage still plays a role (also Simson, 2005b).

City regional approaches to urban forestry have become very common in the United Kingdom, as mentioned by Jones and Baines (2007). All major conurbations in the UK are undergoing major post-industrial regeneration. Significant

Fig. 13.1 In the German Ruhr area, a vast urban forest landscape is emerging, replacing the former industrial landscape (Photo by the author)

urban and community forestry programmes have been a part of this process for the last 20 years. Only more recently, however, are the social, economic and environmental benefits of a substantial 'green infrastructure' more fully recognised in large-scale projects such as the Black Country Urban Forest, Thames Gateway, and the English Community Forests. Simson (2005a, b) also mentions these projects, referring in particular to The White Rose Forest, an initiative that covers the whole West Yorkshire agglomeration. Typical for these types of urban forest landscape projects is that they encompass a strategy for enhancing the tree and woodland cover. They link to regional development and sustainability strategies, using trees and woods as instruments for realising economic, socio-cultural as well as environmental ambitions. These projects act as catalysts for change and stress the importance of partnership working and community involvement. Simon (2001) presents an example from Northern Ireland, the Forest of Belfast partnership. Started in 1992, its strategy is to promote an interest in and appreciation of trees and forestry, as well as to involve the community in tree planting and management. People are encouraged to value trees as integral part of their lives. With the help of funding schemes such as 'Woods on Your Doorstep' (financed by the non-governmental Woodland Trust), new woods have been established in and near Belfast.

Another interesting case of how city forests becomes integrated into a larger-scale, multifunctional landscape is that of the National City Park (NCP) of Stockholm (Fig. 13.2). Schantz (2006) explains how this first urban national park of its kind in the world, which incorporates the Norra Djurgården city forest, was established by parliamentary decision in 1994. The principle of a national city (or urban) park has become incorporated into Swedish environmental and land use legislation, ensuring the area's protection. New built-up areas and installations are only possible under very special conditions, as long as they are not damaging the park landscape, its natural environment, or its natural and cultural values of the historical landscape. The overall aim for the NCP is to act as a model for sustainable development, with focus on nature and ecology, recreation and culture. Swedish legislation stipulates that national urban parks should contain natural areas important for the preservation of urban biodiversity, cultural milieus – including buildings – important for an understanding of national history or that of the city itself, and parks and green areas of architectural or aesthetic significance. National urban parks should be large enough, for example for ecological reasons, but also be an integral part of the urban structure. The Swedish example was followed in Finland, where national city parks – often with city forests as core element – were set up in the cities of Hämeenlinna, Pori and Heinola.

New, larger-scale urban forest landscapes such as the Milan Metrobosco and Parco Nord, the Emscher Landscape Park, the Community Forests of Britain and the Swedish and Finnish national urban parks facilitate planning and management of multifunctional landscapes at the level of the city region. When 'moving up' to higher levels in the planning hierarchy, however, important lessons can be learnt from city forestry. City forest history offers successful examples of how cities and forests are culturally connected, showing that city and forest/nature are

Fig. 13.2 City forests have become part of 'National City Parks', as here in the case of Djurgården in Stockholm, Sweden (Photo by the author)

not just opposites and can join forces in the creation of landscapes with a clear identity of their own.

The term 'green infrastructure' has been mentioned. It comprises an important innovation in the integrative planning of forests and other green spaces. As Longhurst (2007) states, the phrase 'green infrastructure' has become frequently used in countries such as the United Kingdom in reference to urban renaissance and green space regeneration. It can be defined as creating networks of multifunctional green spaces that are carefully planned to meet the environmental, social and economic needs of a community. All green spaces, both public and private, are considered, as the example of London's Green Grid shows (e.g. Worpole, 2006).

The green infrastructure approach brings woodland and other green space to the level of other basic infrastructures of a city region, such as the built and transport infrastructures. Green cities are not merely aesthetically pleasing. They are often also a key engine of economic growth in the modern skills economy (Preston, 2007). Research has shown that green infrastructure planning can add value to the regeneration process. Strategic planning of green infrastructure at the spatial scale of the city region seems the most appropriate, as the regional spatial level is too large to capture important functional linkages and the local authority spatial levels is too small for strategic decisions to be made.

Building Communities

Promoting multifunctional and attractive urban landscapes at the strategic level is one aspect of modern city forestry. Another important aspect is the role of city forests in place making, or perhaps more frequently, as discussed in Chapter 11 'The Social Forest', the re-making of place. As we have seen, place is about close ties between individuals, communities and environments. Place is closely related to home, roots, to identity.

When it comes to place (re-)making, it is important that local communities are increasingly seeking collective improvements to the quality of life ahead or more personal material gains (Smith, 2007a). Citizens increasingly demand to be in the 'driving seat' of shaping their environment (also, Van Herzele et al., 2005; Worpole & Knox, 2007). Policy makers, planners and residents often share a wish for liveable communities, defined as places where citizens are proud to live, are healthy in mind and body and are safe and enjoying the benefits of economic growth.

But how are these liveable communities created (or maintained) and what can be the role of city forests? To start with 'community': as Gilbert (2006) mentions, you can put up and populate buildings – even with green spaces – but you cannot just create a community. What planners and designers cannot plan into space is the meaning that is made up of common memories and shared experiences. Although citizens generally have stronger perceptions of the places where they live, work and relax, the emergence of larger city regions poses an extra challenge to building local identity and commitment (The SAUL Partnership, 2005). Various studies have indicated, however, that it is very important for people's well-being and quality of life to have a shared sense of place at the local level.

Friedman (2000) offers an interesting perspective, writing about the need to balance globalisation and local roots by using the metaphor of the Lexus (a luxurious Japanese car) and the olive tree (a comment element of family land in the Middle East). About olive trees, Friedman (2000, p. 31) writes: "Olive trees are important. They represent everything that roots us, anchors us, identifies us and locates us in the world – whether it be belonging to a family, a community, a tribe, a nation, a religion, or, most of all, a place called home." Friedmann also relates olive trees to, for example, the warmth of the family, to personal rituals and relationships, and to security and feelings of self-esteem. The Lexus, on the other hand, "represents an equally fundamental, age-old human drive – the drive for sustenance, improvement, prosperity and modernization – as it is played out in today's globalization system. The Lexus represents all the burgeoning global markets, financial institutions and computer technologies with which we pursue higher living standards today." (Friedman, 2000, pp. 32–33). The Lexus and the olive tree thus represent the tensions between a quest for material betterment on the one hand, and individual and community identity on the other.

It is tempting to see urban trees as 'literal' olive trees. Trees, city forests and other green spaces play important roles in place making by individuals as well as groups, as we have seen in Chapter 11 'The Social Forest' (Fig. 13.3). City forests

Fig. 13.3 City forests such as Djurgården in Stockholm, Sweden, play an important role in place making. They bring together residents of different walks of life to enjoy a forest landscape (Photo by the author)

have helped provide identities to communities, cities and sometimes even countries. They have been 'sacred' places in the terminology of Tuan (2007) and Kotkin (2005), and subjects of intense conflicts and power struggles. Trees and woodland have also been excellent 'place makers', as symbols and markers of change, longevity, continuity and so forth. Jones and Cloke (2002) mention how, in an increasingly urban and speeded-up world, many organisations and individuals have seized upon the 'time-presence' of trees. In cities dominated by compressed 'time-space', trees and woodland represent ecological timescapes of tree growth and lifespan, interacting with the social. This relates to the need for parochial monuments, landmarks, milestones and other points of reference by which each person can take his or her own bearings in place and time.

Chapter 11 'The Social Forest' showed how physical and social accessibility are crucial in realising the place making potential of city forests and trees (also Worpole & Knox, 2007). Travlou and Thompson (2007) argue that inclusive access to green space can even be regarded a cornerstone of democracy and social equity, and a fundamental condition for social and political participation. Socially-inclusive spaces are accessible open spaces that are not only perceived to belong to everybody, but do in fact attract a high diversity of users (The SAUL Partnership, 2005).

These should ultimately help make people proud of their public space, as well as being involved and interested in it.

As we have seen in Chapter 12 'A Forest of Conflict', city forest history is full of examples of how local society has made a real effort to conserve and develop 'their' forests. This type of commitment is also necessary today, in a time of intense political and economic competition, particularly over issues such as urban land use. Developing the cultural dimension of new urban forest landscapes will require targeting different segments of urban society, including children and youths, the elderly and often marginalised groups such as unemployed, handicapped and ethnic minorities. Many interesting approaches and tools have already been developed for this purpose (e.g. Van Herzele et al., 2005; Janse & Konijnendijk, 2007), but more remains to be done, as community participation is often local and short-term rather than strategic (e.g. Macpherson, 2004). A long-term participation strategy, providing different participatory arenas and settings, is crucial for building of trust and engagement amongst participants in planning processes (Höppner et al., 2007). More also remains to be done for connecting Europe's city forest culture to the urban forest landscapes of the future.

A critical note is needed here. Place making is not easy, and there has been considerable criticism about beliefs that 'sense of community' can be created through the built environment or open space. The general idea is that public space – such as a city forest – plays a special part in providing a venue for chance encounters, for example between segments of the population that hardly meet otherwise. With a higher frequency of social contacts, group formation and social support are strengthened. Talen (1998) recognises the role of the environment in community building, but only with the help of intermediate variables. Non-spatial factors, Talen argues, are very important, including community of interests sought (affiliation with homogeneous, like-minded social group) and avoidance of heterogeneous social interaction. Commonalities such as having children or dogs can facilitate social interaction.

Thus place making and creating joint identity is the result of an interplay between environmental, design and social factors. There is no doubt, however, that social interaction in for example city forests is an important element and should be promoted. Currently, the physical and social accessibility of city forests and other green spaces is not optimal, thus 'locking out' parts of the urban population. Groenewegen et al. (2006) mention how, due to increasing urbanisation combined with a spatial planning policy of densification throughout Europe, more people face the prospect of living in less green residential environments. People without the resources to move to greener areas outside the cities, will be especially affected. This may lead to environmental injustice with regard to the distribution of and access to public green space, as shown in Chapter 11 'The Social Forest'. Limited use of green space can, for example, have negative impacts on public health and well-being, as explained in Chapter 9 'The Healthy Forest'.

Recent surveys in The Netherlands and the United Kingdom suggest a decline in visits to forest and nature, at least amongst certain segments of the population (Bosma, 2002; England leisure visits, 2005). Young people turn more to the popular culture industry and less to traditional culture forms – such as visits to nature areas, including city forests. Ethnic minorities are often not very enthusiastic users of city forests or other nature areas either. In some cases, certain people or groups feel shut out by other people who claim a certain forest or green space as their own – a negative aspect of place making and territoriality. This all means that the city forest becomes, perhaps more than ever, part of an intense competition for people's (leisure) time and money. City forestry needs to enter this competition, as it will be difficult to justify public money being spent on city forest management if nobody is using the forest. Worpole (2006; also Worpole & Knox, 2007) also warns of the risk of people being disconnected from green space. Some people see certain city forests and other green spaces as 'place threatening' rather than place making, for example because they are afraid of crime (see Chapter 3 'The Forest of Fear').

Recently, considerable attention has been given to ways of promoting the physical and social accessibility of city forests and other green space. In Sweden, for example, new manuals provide municipalities and forest owners with advice on how to improve access to, and the use of, city forests and other green space (Bovertket, 2007; Ek, 2007). The importance of informing, as well as of involving and engaging local communities is stressed. Security is a key factor for the use of green space, as also mentioned by Hiss (1990). Referring to experiences from Frankfurt, Hiss states that two basic needs should be met in order to enhance the experience of a public open space. First of all, people need to feel safe and secure and secondly, people need reasons to go to a place – and the more, the better. Adding attractions, 'extra's' and surprises to, for example, a city forest generally enhances its use. As Chapter 7 'A Work of Art' has shown, art can play an important role in creating attractive, surprising city forests. Art can also help in (re-)establishing links between city forests and the local community.

Another way of attracting people to city forests is to concentrate on the health agenda. As mentioned in Chapter 9 'The Healthy Forest', the links between forests, nature and health and well-being have become well established. This offers opportunities for promoting the use of city forests, as healthy lifestyles are coming into fashion. Worpole (2006) offers further advice on how to promote use of green space such as city forests. City forests and other green spaces that have become marginalised for different reasons should be made familiar and loved again. To do so, cultural links can be used, such as local names, associations and local history. The SAUL Partnership (2005) mentions how in new landscape established on abandoned industrial sites, remaining industrial landmarks help to rediscover a landscape's origins in its industrial or cultural past.

Children and youths are a very important group to target in efforts to enhance the use of city forests. Childhood experience and frequency of visits to woods are the most important predictors of how comfortable people feel in the wood and how often they are likely to visit woodland as adults (Milligan & Bingley, 2007).

Fortunately we do have something to build our efforts on. This book has provided ample evidence that close ties between city forests and urban residents do exist. Conflict cases such as those described in Chapter 12 'A Forest of Conflict' show that emotions often run high and that there is sufficient potential for making and re-making place.

Initiatives such as the English Community Forests programme are trying to build on the close ties between cities and forests, even creating new links where no citizen-forest relations exist, for example, because there were no forests before. Community forests are multi-purpose forests that are closely linked to local communities to which they deliver environmental, social and economic benefits (Colangelo et al., 2007). Community forest initiatives in the United Kingdom, Ireland and elsewhere stress the involvement of local residents with the wooded landscape in their living and working environment. As explained by Lambregts and Wiersum (2002), community forestry has traditionally been used in the developing world, with emphasis on the subsistence and direct economic dependence of local communities. In the Western world with its materialistic wealth, emotional ties between forests and local dwellers seem more important. Rather than forest ownership, local commitment is needed. 'Community' has three key dimensions: local territory, social relations and a feeling of connectedness. According to Hiss (1990), connectedness relates to a sense of kinship with all life, as well as a partnership with working landscapes. It has to do with a sense of community and companionability that is traditionally fostered by villages and traditional urban neighbourhoods.

An interesting case of successful place making by local residents is that of a woodland near Camerton in Britain, described in detail by Jones and Cloke (2002). In fact it is better to speak of re-making of place, as the local community redefined their connections to the local landscape. Over time, a woodland was developed on a coal heap. The first trees were planted by the owner of the mine for producing timber as well as for beautification. When the mine was closed during the second half of the 20th century, difficult times arose for the village and its people, who had had the mine as its main employer. However, the woodland started to play an important role in re-establishing community pride and identity. With the planted wood as base, nature started to find its way back into the area, resulting in a gradual and powerful transformation of the landscape. The woodland, which had been appropriated by the local community as a recreational space, started to embody a link to the past as well as to the future. Further greening efforts, such as replanting of parts of the wood by local schools, resulted in a Best-Kept Village award. The forest area became known as 'Little Switzerland' to the residents, although its formal name is now Camerton Batch Local Nature Reserve. There is considerable affection for the trees – even the old, planted coniferous trees are appreciated, as residents compare their scenery as that of old Japanese prints during misty mornings.

Maintaining the Links with Nature

City forests are important parts of multifunctional, attractive urban landscapes. They can make major contributions to providing a sense of place, to place making and re-making. Many of these roles, however, are also provided by city parks and other green spaces, although city forests do often have the advantage of longer historical links to cities and local society. Moreover, in addition to the mentioned roles, city forests also represent a 'wilder' kind of nature. Often they come closest to providing an experience of 'real' nature to urban dwellers in, or very near, their own living environment. Thus city forests are city as well as forest, marketplace as well as privacy reserve, place as well as space. On the one hand they are 'anti-urban' space (Dings & Munk, 2006), yet on the other hand, they are a necessary and integral element of urban life.

The deeper meaning of forest in our cultures has been described earlier, particularly in Chapters 2 'The Spiritual Forest' and Chapter 3 'The Forest of Fear'. Forests continue to play their role as cradle of human society, as 'the other' to civilisation and the cities this has produced. It is important that people can find this 'other', this more natural environment even in cities and towns. This has an impact, as we have seen, on human health and well-being. People become healthier from visiting forests and getting in touch in nature. City forests also provide an escape, a more or less 'safe haven' for those who want to get away from city life, for whatever reason. They allow for 'wilder' behaviour for which there is no room or no social acceptance in more urban parts of the city. Of course this also creates problems of its own which have to be managed. Still, as we have seen in this book, cities need their fringe areas.

City forests are very important for shaping our overall relationships with, and views of nature, as nature views today are largely created by urban society and thus dominated by urban values. In a largely artificial environment, city forests leave room for natural processes, the changing of seasons and the passage of time. They harbour a certain amount of mystery, evoking excitement and perhaps even fear. City forests are also forests of learning, showing urban dwellers, and especially children and youths, what nature is all about. They link us to our rural past, when the forest was a crucial provider of products.

In Chapter 8 'The Wild Side of Town', the example of Zurich's Sihlwald was given (Fig. 13.4). This case demonstrates how decision-makers currently emphasise the ecological aspects of city forests, combining nature preservation – or, as in the case of countries such as The Netherlands, nature development – with recreation and education. The 'wilderness' is brought back into the city – the city that once destroyed it. In some cases, nature itself does the work, reclaiming abandoned industrial and other sites. In Germany as well as in other countries, for example, a new type of city forest is emerging, coined 'urban-industrial woodland' by Kowarik (2005). These woodlands are heavily culturally influenced, but still represent a natural development that occurs independently on typical urban-industrial sites.

In the case of the Emscher Landscape Park in the German Ruhr area, the new 'wild forests' in the middle of the city are part of an effort to revitalise the area and create new links between people and nature. A central concept is that of 'Natur auf Zeit' (nature on time) which refers to the temporary character of some of the new nature areas. Conversion to nature such as forest, be it on a shorter or longer term, takes the pressure away from land owners with regard to compensation obligations for land areas that temporarily become part of the 'industrial forest'. In the Emscher Landscape Park, nature adventure is an important part of the new leisure activities offered on everybody's doorstep. Wild nature becomes part of a 'healing

Fig. 13.4 The Sihlwald near Zurich is an example of how wilderness experiences are brought back to the city (Photo by the author)

process' – spaces with a new, peculiar aesthetics and particular atmosphere are replacing areas that previously appeared derelict and rather intimidating (Landesbetrieb Wald und Holz NRW, 2006; Neiss, 2007).

City forests, old and new, offer a much-needed supplement and perhaps even contrast to the more 'artificial' representations of nature that tend to dominate our cities and towns. One can think here of the highly-designed city park, private gardens, and zoological gardens and theme parks. Davis (1996) argues how in theme parks such as Sea World, an idealised form of nature is 'designed' and presented as magical. Stories of nature are carefully crafted, often targeted towards certain audiences. There is a risk that this de-contextualised, 'idealised' nature gives a wrong impression of what nature is like, colouring people's views of nature. Davis (1996, p. 217) also sees a risk in the 'privatisation' of nature that is embodied in the work of large theme parks: "Definitions of nature and the solutions to its problems are now massively authored by the private sphere of conglomerate, corporate culture, at the same time that corporations claim to further the public good."

The risks of a distorted nature view are also touched upon by Lawrence (2006), who mentions how some trees or tree places have become 'museum pieces', considered as being static, but still harbouring living organisms and being in need of management. This type of attitude explains why Rackham (2004, p. 207) writes, when listing the ingredients of a good management plan for a woodland: "Conservation is about letting trees be trees, not gateposts with leaves".

Final Thoughts

City forests have a long tradition and rich meaning, in Europe, but also in other parts of the world. Even those city forests that are centuries old are often still meeting the demands of modern urban society. They are appreciated for contributing to a pleasant living, working and leisure environment, making people proud of their local environment, and for bringing nature to their doorstep. In this way, city forests have been at the forefront of a re-appreciation of the important role of forests and nature in an urbanised society.

At the outset of this book, the concept of 'forest' was introduced, as was its changing meaning over time. City forestry and related approaches such as urban forestry may well play a role in the further evolution of what it considered to be a 'forest'. As Clark and Jauiainen (2006) write, urban society encompasses a constant adaptation, refining and reformulation in time and space of ideas and policies. In the Community Forests of the UK and Ireland and the Parco Nord of Milan, for example, 'community forest' is envisaged as a rather open, multifunctional landscape with a tree cover of perhaps 30–40%. The planning of new city forests in Flanders, Belgium, has also led to a discussion about what a 'city forest' really is (Van Herzele, 2005). This chapter has described how city forests have become part of more strategic landscape and land use planning. City forests are important elements of urban forest landscapes, of urban green structures, of cultural urban landscapes, and of sustainable and competitive city regions.

City forestry has also influenced the development of forestry at large, as several authors have acknowledged (e.g. Krott, 1998, 2004; Otto 1998; Konijnendijk, 2003; Nowak et al., 2005). As shown in Chapter 10 'The Forest of Learning', city forests have offered a fruitful testing grounds for introducing new concepts and management, focusing, for example, more on social and environmental services, forestry under anthropogenic and other stresses, and on the involvement of stakeholders and on conflict management. Innovation continues, with landscape laboratories, school forests and play forests, GPS walks, urban national parks and wild industrial woodland. Forestry has slowly started to accept and even embrace its urban mandate (Essmann et al., 2007; Pröbstl, 2007). Urbanisation and sprawl are impacting on larger areas of forests, offering forestry, as Essmann et al. (2007, p. 61) write, "a challenge outside its traditional rural domain". In particular, the use and management of those forests in and closest to urban areas have changed. Issues such as fires at the urban-wildland interface, the occurrence of exotic pests and forest fragmentation have become very important (Nowak et al., 2005).

But even more important is that forestry at large is increasingly dominated by urban societies and their values, norms and demands. In this way, Nowak et al. (2005) write, all forestry is becoming urban in the sense that it must respond to urban perceptions and needs. Moreover, as in the case of city forests, multiple use forestry is being transformed into modern landscape planning. This type of planning extends across the diverse ownerships and jurisdictions, is consistent with sustainability, and carried out jointly with mutual participation of landowners and managers, stakeholders, users and public interests, in the determination of the preferred landscape long-term outcome (Fedkiw, 2007). This also means that forestry has to engage even more than before with other fields, such as landscape planning, landscape architecture, ecology, but also political science, the social sciences and economics. Urban forestry offers excellent opportunities for this dialogue between different fields, as well as for a wider public discussion about forests and forestry (Nowak et al., 2005).

In several highly urbanised countries, experiences from city forestry and urban forestry have come to influence general forest policy. This is nicely illustrated by the new strategy for England's trees, woods and forests (DEFRA, 2007), which has 'communities and places' as one of its three central themes. The strategy looks at how forests can contribute to cohesive and engaged communities, sense of place and green infrastructure. Among the specific topics addressed are public health, safety, community involvement, improved access and the creation of liveable neighbourhoods.

It is good and necessary that city forests are increasingly seen as part of a bigger picture. Yet, the city forest heritage also needs to be cherished, recognising each individual city forest as a unique landscape, and for the many social, cultural and other roles it has played. This does not mean that all existing city forests should be preserved in their present state. Although their cultural-historical values do require recognition and conservation, having city forests in rapidly changing and demanding urban settings can only be justified if these woodlands cater for the needs of modern society. Once these needs related to timber for building houses and ships, to wood for warming houses and powering industries, and to providing food. Later, city forests brought prestige and even identity, offered settings for recreation, and helped protect our drinking water resources. Today, city forests have become appreciated as well as therapeutic landscapes and as learning forests that keep urban populations in touch with nature. These many important contributions of city forests to urban society should all be kept in mind when we maintain and develop city forests as true symbiosis of forest and city – our Forests for the Future.

Literature

Aagaard, M. (2007, August 1). Polsk urskov får EU til at se rødt. *Politiken*, 7 (in Danish).
Aalde, H. (1992). A small case-study on the forest areas surrounding Oslo. In: J. Hummel, & M.P.E. Parren (Eds.), *Forests, a growing concern. Proceedings of the XIXth IFSS, Wageningen, September 30–October 7, 1991* (pp. 43–47). Gland/Cambridge: IUCN.
Agentschap voor Bos en Natuur. (2006). *Doorblader het bos. Week van het Bos, 1–8 oktober 2006.* Campaign newspaper. Brussels: VBV (in Dutch).
A history of the county of Middlesex. (1989). *Hampstead: Hampstead Heath.* In: Volume 9, Hampstead and Paddington, pp. 75–81. Retrieved July 30, 2007 from British History Online Web site: http://www.british-history.ac.uk/report.asp?compid = 22644.
Akbari, H., Davis, S., Dorsano, S., Huang, J., & Winert, S. (1992). *Cooling our communities: A guidebook on tree planting and white coloured surfacing.* Washington DC: US Environmental Protection Agency, Office of Policy Analysis, Climate Change Division.
Alkmaarder Houte dupe van onjuist beheer in verleden. (1992). *Boomblad* 1992 (December), 4–7 (in Dutch).
Alleen tussen de frontlinies van Sarajevo is nog hout. (1994, December 8). *De Volkskrant*, p. 5 (in Dutch).
Allender, S., & Rayner, M. (2007). The burden of overweight and obesity-related ill health in the UK. *Obesity Reviews* 8(5), 467–473.
Allman, T.D. (2007). The theme-parking, megachurching, franchising, exurbing, McMansioning of America; How Walt Disney changes everything. *National Geographic* 2007(3), 96–115.
Alvey, A.A. (2006). Promoting and preserving biodiversity in the urban forest. *Urban Forestry & Urban Greening* 5(4), 195–201.
Amati, M., & Yokohari, M. (2006). Temporal changes and local variations in the functions of London's Green Belt. *Landscape and Urban Planning* 75, 125–142.
Amati, M., & Yokohari, M. (2007). The establishment of the London Greenbelt: reaching consensus over purchasing land. *Journal of Planning History* 2007(6), 311–339.
Ammer, U., Weidenbach, M., Beer, M., & Hwang, Y.-H. (1999). *Landschafts- und erholungsplanerische Entwicklungsstudie für die Wildparke im Ebersberger Forst und im Forstenrieder Park.* Retrieved July 26, 2007 from Landconsult Web site: http://landconsult.de/markus/wildpark (in German).
Anan'ich, B., & Kobak, A. (2006). St Petersburg and urban green space, 1850–2000: an introduction. In: P. Clark (Ed.), *The European city and green space. London, Stockholm, Helsinki and St Petersburg, 1850–2000* (pp. 247–271). Hants: Ashgate Historical Urban Studies.
Andela, G. (1996). De gemaakte natuur. In: H. Moscoviter (Ed.), *Op de groei gemaakt. 'Geriefelijkheden voor een wel-ingerichte stad'* (pp. 104–129). Rotterdam: Gemeentewerken (in Dutch).
Anema, K. (1999). *Het Haagse Bos.* Abcoude: Staatsbosbeheer and Uitgeverij Uniepers (in Dutch).
Appleton, J. (1996). *The experience of landscape.* 2nd edition. Chichester: Wiley.

A queer crime in the city parks, and queer pleasures in high society. (1901, July 8). *New York Times*. Retrieved December 4, 2007 from New York Times Archives Web site: http://query.nytimes.com/gst/abstract.html?res = 9802EFDF163BE733A2575BC0A9619C946097D6CF.

Ardoin, N.M. (2004). Sense of place and environmentally responsible behaviour: what the research says. In: *Proceedings of Biloxi conference, 2004*. Retrieved July 12, 2006 from North American Association for Environmental Education Web site: http://naaee.org/conferences/biloxi/n_ardoin_3_10008a.pdf.

Århus Kommune. (2004). *På tur I Riis skov*. Århus: Naturforvaltningen, Århus Kommune (in Danish).

Arnberger, A., Brandenburg, C., & Eder, R. (2005). Dog walkers as management challenge in urban forests. Abstract. In: P. Neuhöfervá (Ed.), *Management of urban forests around large cities. Proceedings of abstracts, Prague, October 3–5, 2005* (p. 15). Prague: Faculty of Forestry and Environment, Czech University of Agriculture.

Arnberger, A., & Eder, R. (2007). Visitor satisfaction with an artificial urban forest in Vienna. In: *Abstracts, 'New forests after old industries', 10th European Forum on Urban Forestry, May 16–19, 2007, Gelsenkirchen* (pp. 58–59). Dortmund: Landesbetrieb Wald und Holz Nordrhein-Westfalen.

Arnold, K.R. (1967). Scope of recreation research. In: *Proceedings of the XIV IUFRO World Congress, München 1967, Papers VII, section 26* (pp. 1–13). Vienna: IUFRO.

Arntzen, S. (2002). Cultural landscapes and approaches to nature – Ecophilosophical perspectives. In: V. Sarapik, K. Tüür, & M. Laanemates (Eds.), *Place & Location II. Proceedings of the Estonian Academy of Arts 10* (pp. 27–50). Tallinn: Eesti Kunstakadeemia.

Australia eyes pill for kangaroos. (2006, August 23). Retrieved August 12, 2007 from BBC News Web site: http://news.bbc.co.uk/2/hi/asia-pacific/5278010.stm.

Avdibegović, M. (2003). Program for protection of Bosna river source – the importance of the area from a cultural perspective. Unpublished report. Sarejevo: Faculty of Forestry, University of Sarajevo.

Baaij, G. (2002). 111 ontwerpen voor het Museumbos. *Nederlands Bosbouwtijdschrift* 74, 25–27 (in Dutch).

Bachmann, P. (2006). Neue Pärke für die Schweiz. *GeoAgenda* 2006(6), 4–7 (in German).

Bader, G. (2007). Basel-Stadt und seine Wälder. *Schweizerische Zeitschrift für Forstwesen* 158(7), 216–220 (in German, with English abstract).

Baeté, H. (2006). Zwijnerij in Bos ter Rijst. *Bosrevue* 16, 12 (in Dutch).

Baines, C. (1986). *The wild side of town*. How you can help and enjoy the wildlife around you. London: BBC Publications and Elm Tree Books.

Bakri, N. (2005, December 28). Beirut's Pine Forest withering away. *The Daily Star*.

Balk, T. (1979). *Een kruiwagen vol bomen: verleden en heden van het Amsterdamse Bos*. Amsterdam: Staatsdrukkerij (in Dutch).

Ballik, K. (1993). Wiener Erholungs- und Quellenschutzwälder. *Österreichische Forstzeitung* 104(8), 8–10 (in German).

Barbosa, O., Tratalos, J.A., Armsworth, P.R., Davies, R.G., Fuller, R.A., Johnson, P., & Gaston, K.J. (2007). Who benefits from access to green space? A case study from Sheffield, UK. *Landscape and Urban Planning* 82(2–3), 187–195.

Barr, B.M., & Braden, K.E. (1988). *The disappearing Russian forest*. Totowa: Rowman and Littlefield.

Bell, S., Blom, D., Rautamäki, M., Castel-Branco, C., Simson, A., & Olsen, I.A. (2005). Design of urban forests. In: C.C. Konijnendijk, K. Nilsson, T.B. Randrup, & J. Schipperijn (Eds.), *Urban forests and trees – a reference book* (pp. 149–186). Berlin: Springer.

Bell, S., Thompson, C.W., & Travlou, P. (2004). Contested views of freedom and control: children, teenagers and urban fringe woodlands in Central Scotland. *Urban Forestry & Urban Greening* 2(2), 87–100.

Berglund, E. (2007). The fear factor. *Green Places* 2007(38), 24–25.

Berliner Forsten. (2001). *Vom Kulturwald zum Naturwald. Landschaftspflegekonzept Grunewald*. Arbeitsmaterialen der Berliner Forsten 1. Berlin: Senatsverwaltung für Stadtentwicklung und Umweltschutz (in German).

Berliner Forsten. (2006). Retrieved August 3, 2006 from Berliner Forsten Web site: http://www.stadtentwicklung-berlin.de/forsten (in German).

Birnie, R. (2007). Rural Landscapes: for food or fun? *Green Places* 35(2007), 28–31.

Bisenieks, J. (2001). Management of urban forests of Latvia. In: C.C. Konijnendijk, & Flemish Forest Organisation (Eds.), *Communicating and financing urban forests. Proceedings of the 2nd and 3rd IUFRO European Forum on Urban Forestry, Aarhus (May 1999) and Budapest (May 2000)* (pp. 95–99). Brussels: Ministerie van de Vlaamse Gemeenschap, Afdeling Bos en Groen.

Bitterlich, W. (1967). Allgemeine Wohlfahrtsbewertung siedlungsnaher Wälder. In: *Proceedings of XIV. IUFRO Congress, München, 1967, VII, Section 26* (pp. 406–411). Vienna: IUFRO (in German).

Blok, E. (1994). Bomen door de eeuwen heen. *Bomennieuws* 1994(4), 102–105 (in Dutch).

Blom, D. (2005). The design of urban woodlands in the Netherlands: development of 'polder forests'. In: C.C. Konijnendijk, J. Schipperijn, & K. Nilsson (Eds.), *COST Action E12: Urban forests and trees – Proceedings No 2* (pp. 65–88). Luxembourg: Office for Official Publications of the European Communities.

Böttcher, S. (2001). *Groene en recreatieve bedrijventerreinen; de houding van managers van bedrijven.* Internal report. Wageningen: Alterra, Research Instituut voor de Groene Ruimte (in Dutch).

Bonnaire, E. (1992). Urban forestry and foresters in France. In: J.A. Hummel, & M.P.E. Parren (Eds.), *Forests, a growing concern. Proceedings of the XIXth IFSS, Wageningen, September 30–October 7, 1991* (pp. 38–42). Gland/Cambridge: IUCN.

Bonnes, M., Carrus, G., Bonaiuto, M., Fornara, F., & Passafaro, P. (2004). Inhabitants? Environmental perceptions in the city of Rome within the UNESCO Programme on Man and Biosphere Framework for Urban Biosphere Reserves. *Annals of the New York Academy of Sciences* 1023, 1–12.

Borgemeister, B. (2005). *Die Stadt und ihr Wald – Eine Untersuchung zur Waldgeschichte det Städte Göttingen und Hannover vom 13. bis zum 18. Jahrhundert.* Hannover: Hahnsche Buchhandlung (in German).

Bos, R. (1994, March 26). Een ommetje door de Haarlemmer Hout. *De Volkskrant* (in Dutch).

Bosma, N. (2002). Vraagtekens bij het bezoek aan bos- en natuurgebieden. *Nederlands Bosbouwtijdschrift* 74, 22–24 (in Dutch).

Bosma, N.J., & Gaasbeek, N.H. (1989). *Noorwegen: verslag van een bosbouwstage.* Wageningen: Vakgroep Bosbouw, Landbouwuniversiteit (in Dutch).

Boutefeu, B. (2007). *La forêt comme un théâtre ou les conditions d'une misse en scène réussi.* Doctoral dissertation. Lyon: École Normale Supérieure Lettres et Sciences Humaines and ONF (in French).

Boverket. (2007). *Bostadsnära natur – inspiration & vägledning.* Karlskrona: Boverket (in Swedish).

Brandl, H. (1985). Aus der Geschichte der städtischen Forstverwaltung Freiburg. *Allgemeine Forstzeitschrift* 37, 950–954 (in German).

Bregman, H. (1991). Het Amsterdamse Bos. *Nederlands Bosbouwtijdschrift* 63, 94–99 (in Dutch).

Brésard, C.R. (1995). Passions et délits à Fontainebleau. *Le Courrier de la Nature* no. 150, 7–9 (in French).

Brink, N. (1984). *Haarlemmerhout 400 jaar. "Mooier is de natuur nergens".* Haarlem: Schuyt (in Dutch).

Bryson, B. (2007). *The life and times of the Thunderbolt Kid – Travels through my childhood.* London: Transworld Publishers.

Buchinger, M. (1967). Forest recreation versus conservation. History, legislation, management and research. In: *Proceedings of IUFRO Congress, München, 1967, VII, Section 26* (pp. 265–285). Vienna: IUFRO.

Bucht, E. (2002). Traditions in urban park planning and management in Sweden and other European countries. In: T.B. Randrup, C.C. Konijnendijk, T. Christophersen, & K. Nilsson (Eds.), *COST Action E12: Urban Forests and Trees – Proceedings No. 1* (pp. 215–226). Luxembourg: Office for Official Publications of the European Communities.

Bürg, J., Ottitsch, A., & Pregernig, M. (1999). *Die Wiener und ihre Wälder. Zusammenfassende Analyse sozioökonomischer Erhebungen über die Beziehung der Wiener Stadtbevölkerung zu Wald und Walderholung*. Schriftenreihe des Instituts für Sozioökonomik der Forst- und Holzwirtschaft Band 37. Wien: Universität für Bodenkultur (in German).

Bues, C.-Th., & Triebel, J. (2004). Urbane Forstwirtschaft – eine neue Herausforderung. *Stadt + Grün* 2004(4), 37–41 (in German).

Buijs, A.E., Langers, F., & De Vries, S. (2006). *Een andere kijk op groen. Beleving van natuur en landschap in Nederland door allochtonen en jongeren*. Rapport 24, Wettelijke Onderzoekstaken Natuur & Milieu. Wageningen: Alterra (in Dutch).

Buis, J. (1985). *Historia Forestis: Nederlandse bosgeschiedenis*. Two parts. Utrecht: HES Uitgevers (in Dutch).

BUND Berlin. (2007). *Wildnis in Berlin*. Retrieved July 24 from BUND Berlin Web site: http://www.wildnis-in-berlin.de (in German).

Burgers, J. (2000). Urban landscapes: on public space in the post-industrial city. *Journal of Housing and the Built Environment* 15, 145–164.

Burgess, J. (1995). *'Growing in confidence' – Understanding people's perceptions of urban fringe woodlands*. Technical Report. Cheltenham: Countryside Commission.

Burkhardt, I., Schober, F., & Pauleit, S. (2007). Ecological urban regeneration through the establishment of urban woodlands in inner-city areas – a contribution to sustainable urban development in the City of Leipzig. In: *Abstracts, 'New forests after old industries'. 10th European Forum on Urban Forestry, May 16–19, 2007, Gelsenkirchen* (pp. 52). Dortmund: Landesbetrieb Wald und Holz Nordrhein-Westfalen.

Byrnie, R. (2007). Rural landscapes: for food or fun? *Green Places* 2007(35), 28–31.

CABE Space. (2005). *Does money grow on trees?* London: Commission for Architecture and the Built Environment.

Caldecott, M. (1993). *Myths of the sacred tree*. Rochester: Destiny Books.

Caspers, T. (1999). *Het Mastbos*. Abcoude: Staatsbosbeheer & Uitgeverij Uniepers (in Dutch).

Cheng, A.S., Kruger, L.E., & Daviels, S.E. (2003). 'Place' as an integrating concept in natural resource politics: propositions for a social science research agenda. *Society and Natural Resources* 16, 87–104.

Chiesura, A. (2004). The role of urban parks for a sustainable city. *Landscape and Urban Planning* 68(2004), 129–138.

Cieslak, E., & Biernat, C. (1995). *History of Gdansk*. Translated by B. Blaim, & G.M. Hyde. Gdansk.

City of Helsinki. (2005). *Nature, recreation and sports in Helsinki*. Urban Facts. City of Helsinki.

City of Konstanz. (2007). On-site information panels in the Lorettowald (visited July 20, 2007).

City of London Corporation. (1998). *Epping Forest Management Plan. April 1998 to March 2003*. Summary document. London: City of London Corporation.

City of London Corporation. (2007). Highgate Wood. Information leaflet. Retrieved August 3, 2007 from City of London Web site: http://www.cityoflondon.gov.uk/Corporation/living_environment/open_spaces/highgate_wood.htm.

Clark, P., & Hietala, M. (2006). Helsinki and green space, 1850–2000: an introduction. In: P. Clark (Ed.), *The European city and green space. London, Stockholm, Helsinki and St Petersburg, 1850–2000* (pp. 175–187). Hants: Ashgate Historical Urban Studies.

Clark, P., & Jauhiainen, J.S. (2006). Introduction. In: P. Clark (Ed.), *The European City and Green Space. London, Stockholm, Helsinki and St Petersburg, 1850–2000* (pp. 1–29). Hants: Ashgate Historical Urban Studies.

Clifford, S. (1994). Trees, woods, culture and imagination. Creating meaning for people in new forests. In: K. Chambers, & M. Sangster (Eds.), *A Seed in Time. Proceedings of the Third International Conference on Urban and Community Forests, Manchester, August 31–September 2, 1993* (pp. 11–18). Edinburgh: Forestry Commission.

Cnudde, J. (2007). Boompjesweekend 2007: koop een boom, plant een bos! *De Boskrant* 37(1), 4–5 (in Dutch).

Coatney, L.R. (1993). *The Katyn Massacre: An assessment of its significance as a public and historical issue in the United States and Great Britain, 1940–1993*. Master thesis. Western Illinois

University, Department of History. Retrieved December 3, 2007 from IBiblio Web site: http://www.ibiblio.org/pub/academic/history/marshall/military/wwii/special.studies/katyn.massacre/katynlrc.txt.

Colangelo, G., Fiore, M., Davies, C., Sanesi, G., & Lafortezza, R. (2007). The role of community forests in the process of rehabilitating decommissioned industrial sites. In: *Abstracts, 'New forests after old industries'. 10th European Forum on Urban Forestry, May 16–19, 2007, Gelsenkirchen* (pp. 37–39). Dortmund: Landesbetrieb Wald und Holz Nordrhein-Westfalen.

Coles, R.W., & Bussey, S.C. (2000). Urban forest landscapes in the UK – progressing the social agenda. *Landscape and Urban Planning* 52(2000), 181–188.

Common Ground. (2007). *Mission statement.* Retrieved August 5, 2006 from Common Ground Web site: http://www.commonground.org.uk.

Comrie, A. (2007). Climate change and human health. *Geography Compass* 1(3), 325–339.

Connaissez-vous les résidences présidentielles? (2005, July). Retrieved December 5, 2007 from L'Internaute Magazine Web site: http://www.linternaute.com/actualite/dossier/05/vacances-politiques/les-residences-presidentielles.shtml (in French).

Conway, T., & Urbani, L. (2007). Variations in municipal urban forestry policies: a case study of Toronto, Canada. *Urban Forestry & Urban Greening* 6(3), 181–192.

Cooper, J., & Collinson, N. (2006). Space for people: targetting action for woodland access. In: F. Ferrini, F. Salbitano, & G. Sanesi (Eds.), *Urban forestry: Bridging cultures, disciplines, old attitudes and new demands. 9th European Forum on Urban Forestry, Florence, May 22–26, 2006* (pp. 71–74). Abstract book. Florence: University of Florence.

Cordall, L. (1998). *The role of the landscape architect in the rebuilding of a war-torn city. Sarajevo compared with Conventry and Beirut.* Retrieved August 7, 2006 from Friends & Partners Web site: http://www.friends-partners.org/bosnia/cord1.html.

Cornelis, J., & Hermy, M. (2004). Biodiversity relationships in urban and suburban parks in Flanders. *Landscape and Urban Planning* 69, 385–401.

Cornelius, R. (1995). *Geschichte der Waldentwicklung: die Veränderung der Wälder durch die Waldnutzungen und Immissionsbelastungen seit dem Mittelalter.* Monitoring Programm Naturhaushalt Heft 3. Berlin: Senatsverwaltung für Stadtentwicklung und Umweltschutz (in German).

Corvol, A. (1991). Le bois et la ville du moyen age au XXe siècle. Paris: Éditions ENS, Hors collections des Cahiers de Fontenay (in French).

Crampton, J.W., & Elden, S. (2006). *Space, knowledge and power: Foucault and geography. Introduction* (pp. 1–14). Williston: Ashgate.

Créton, R. (2006, September 17). Zweedse strijd om wolven. *De Telegraaf*, T6 (in Dutch).

Crews, J. (2003). Forest and tree symbolism in folklore. *Unasylva* 54(2). Retrieved August 9, 2007 from FAO Unasylva Web site: http://www.fao.org/forestry/site/unasylva/en/.

Cronon, W. (1996a). Introduction: in search of nature. In: W. Cronon (Ed.), *Uncommon ground. Rethinking the human place in nature* (pp. 23–56). New York & London: W.W. Norton & Company.

Cronon, W. (1996b). The trouble with wilderness; or, getting back to the wrong nature. In: W. Cronon (Ed.), *Uncommon ground. Rethinking the human place in nature* (pp. 69–90). New York & London: W.W. Norton & Company.

Cronon, W. (Ed.). (1996c). *Uncommon ground. Rethinking the human place in nature* (p. 489). New York & London: W.W. Norton & Company.

Dalum, M. (2002, May 19). Slaget om Brøndbyskoven. *Politiken*, pp. 1–2 (in Danish).

Danmarks Naturfredningsforening and TNS Gallup. (2004). Børn og Natur. Copenhagen: DNF (in Danish).

Darge, P. (2006). Van bomen en jongensdromen. *Weekend Knack* 2006(30), 26–29 (in Dutch).

Davies, C., & Vaughan, J. (2001). It's my forest – Creating common ownership and care of the urban forest. In: C.C. Konijnendijk, & Flemish Forest Organisation (Eds.), *Communicating and financing urban forests. Proceedings of the 2nd and 3rd IUFRO European Forum on Urban Forestry, Aarhus (May 1999) and Budapest (May 2000)* (pp. 35–47). Brussels: Afdeling Bos en Groen, Ministerie van de Vlaamse Gemeenschap.

Davis, B. (2007). NYSAF hosts foresters from the Republic of Armenia. *The Forestry Source* 2007(May), 8.

Davis, S.G. (1996). "Touch the magic". In: W. Cronon (Ed.), *Uncommon ground. Rethinking the human place in nature* (pp. 204–217). New York & London: W.W. Norton & Company.

De eerste echte Lappersfortbaby is er binnenkort – Bewoners delen al jarenlang lief en leed in de bomen. (2002, September 21). *Gazet van Antwerpen* (in Dutch).

Deforche, B.I., De Bourdeauduij, I.M., & Tanghe, A.P. (2006). Attitude towards physical activity in normal-weight, overweight and obese adolescents. *Journal of Adolescent Health* 38(5), 560–568.

DEFRA (Department for Environment, Food and Rural Affairs). (2007). A strategy for England's trees, woods and forests. London.

Den her skov er alligevel så røvkedelig. (2006, July 18). *Politiken* (in Danish).

Dennis, S.F. (2006). Nearby nature. The new Pritzker Family Children's Zoo introduces native species to (sub)urban kids. *Landscape Architecture* 2006(November), 30–41.

Den Ouden, J., & De Baaij, G. (2005). *Bos; de verzekering voor je gezondheid.* 4e Nationale Bosdebat, 18 mei 2005, te Velp. Retrieved July 26, 2007 from KNBV Web site: http://www.knbv.nl/archief/3/bos-de-verzekering-voor-je-gezondheid (in Dutch).

Derex, J.-M. (1997a). *Histoire du Bois de Boulogne. Le bois du roi et la promenade mondaine de Paris.* Paris: L'Harmattan (in French).

Derex, J.-M. (1997b). *Histoire du Bois de Vincennes. La forêt du roi et le bois du people de Paris.* Paris: L'Harmattan (in French).

De Vreese, R., & Van Nevel, L. (2006). Een speelbos realiseren, gemakkelijker dan je denkt! *Bosrevue* 18, 7–10 (in Dutch).

De Vries, S., & De Bruin, A.H. (1998). Segmenting recreationists on the basis of constraints. Report 121. DLO Winand Staring Centre, Wageningen.

De Vries, S., Verheij, R.A., Groenewegen, P.P., & Spreeuwenberg, P. (2003). Natural environments – healthy environments? An exploratory analysis of the relationship between green space and health. *Environment and Planning A* 35, 1717–1731.

Di Chiro, G. (1996). Nature as community: the convergence of environment and social justice. In: W. Cronon (Ed.), *Uncommon ground. Rethinking the human place in nature* (pp. 298–320). New York & London: W.W. Norton & Company.

Dienst Amsterdam Beheer. (2002). *Een Bos voor heel Amsterdam. Concept Beleidsplan Amsterdamse Bos 2002–2010.* Amsterdam: Dienst Amsterdam Beheer (in Dutch).

Dietvorst, A. (1995). Het landschap van morgen. *Groen* 1995(12), 9–10 (in Dutch).

Dings, M., & Munk, K. (2006). Parkgeluk. *HP De Tijd*, May 12, 2006, 42–49 (in Dutch).

District Valuer's Report for the Forestry Commission on Bold Colliery Power Station Site. (2004). Report issued February 2004. Edinburgh: Forestry Commission.

Dollar, D. (2006). China's golden cities. *Newsweek*, 2006(July 3–10), 63.

Dragt, G. (1996). Het Valkenberg te Breda: van bos en hoftuin tot stadspark. *Groen* 1996(1), 9–12 (in Dutch).

Dua heeft voorakkoord met Fabricom over Lappersfortbos. (2003, April 18). *De Financieel-Economische Tijd* (in Dutch).

Dubrovnik menaced by major fire. (2007, August 6). Retrieved August 6, 2007 from BBC News Web site: http://news.bbc.co.uk/.

Duinwaterbedrijf Zuid-Holland. (2007). *Geniet van de duinen.* Retrieved October 12, 2007 from Duinwaterbedrijf Zuid-Holland Web site: http://www.dzh.nl/renderer.do/clearState/true/menuId/59707/returnPage/59706/ (in Dutch).

Dwyer, J., Schroeder, H.W., & Gobster, P. (1991). The significance of urban trees and forests: towards a deeper understanding of values. *Journal of Arboriculture* 17(10), 276–284.

Eelde zit met slangen in de maag. (2006, July 28). *De Telegraaf.* Retrieved July 28, 2006 from De Telegraaf Web site: http://www.telegraaf.nl/binnenland/47300891/Eelde_zit_met_slangen_in_de_maag.html (in Dutch).

Ek, B. (2007). Få ut flere i skogen! Får man vara med att planera och bestämma ökar intresset. *Skogen* 2007(6–7), 44–45 (in Swedish).

Eker, Ö., & Ok, K. (2005). Results of changing social demands in Istanbul Forest Enterprise: a case study. In: C.C. Konijnendijk, J. Schipperijn, & K. Nilsson (Eds.), *COST Action E12: Urban forests and trees – Proceedings No 2* (pp. 119–131). Luxembourg: Office for Official Publications of the European Communities.

Ellefson, P.V. (1992). *Forest resources policy: Process, participants, and programs*. New York: McGraw-Hill.

Elsasser, P. (1994). *Waldbesuch in Hamburg – Ergebnisse einer Bürgerbefragung im Hamburger Stadtgebiet*. Arbeitsbericht des Instituts für Ökonomi 94/4. Hamburg: Bundesforschungsanstalt für Forst- und Holzwirtschaft (in German).

England leisure visits. Report of the 2005 survey. (2005). London: Natural England. Retrieved December 5, 2007 from: http://www.countryside.gov.uk/LAR/Recreation/visits/index.asp

Er was eens, toen de dieren nog spraken en er geen TV was… (2006, August 11). *De Volkskrant* (in Dutch).

Essmann, H.F., Andrian, G., Pettenella, D., & Vantomme, P. (2007). Influence of globalization on forests and forestry. *Allgemeine Forst und Jagdzeitung* 178(4), 59–68.

European Landscape Convention. (2000). Retrieved October 29, 2007 from Council of Europe Web site: http://www.coe.int/t/e/Cultural_Co-operation/Environment/Landscape/.

European Science Foundation. (2004). *European Science Foundation Workshop – Urban Civilization: Culture Meets Commerce, Prague, September 17–18, 2004*. Report. Brussels: European Science Foundation.

European Union. (2007). *Lisbon Strategy*. Retrieved November 16, 2007 from European Union Web site: http://europa.eu/scadplus/glossary/lisbon_strategy_en.htm.

FACE (International Association of National Hunters' Associations). (2007). *Census of the number of hunters in Europe* (2007). Retrieved August 11, 2007 from the FACE Web site: http://www.face.eu/fs-hunting.htm.

Fachbereich Umwelt und Stadtgrün. (2004). *Stadtwälder in Hannover – Die Eilenriede*. Hannover (in German).

Fairbrother, N. (1972). *New lives, new landscapes*. 2nd edition. Ringwood: Pelican Books.

FAO. (1997). *Issues and opportunities in the evolution of private forestry and forestry extension in several countries with economies in transition in Central and Eastern Europe. Poland*. Retrieved August 18, 2006 from FAO Web site: http://www.fao.org/docrep/w7170E/w7170e0e.htm.

Fariello, F. (1985). *Architettura dei Giardini*. Roma: Edizione dell'Ateneo (in Italian).

Farjon, J.M.J., Hazendonk, N.F.C., & Hoeffnagel, W.J.C. (1997). *Verkenning natuur en verstedelijking 1995–2020*. Wageningen: IKC Natuurbeheer (in Dutch).

Fedkiw, J. (2007). Where did multiple use go? *Journal of Forestry* 105(4), 213–214.

Fei, S. (2007). The Geography of American Place Names and Trees. *Journal of Forestry* 105, 84–90.

Ferber, U., Grimski, D., Millar, K., & Nathanail, P. (Eds.). (2006). *Sustainable brownfield regeneration: CABERNET network report*. Nottingham: University of Nottingham.

Fernandes, C. (2007, June 22). *Malaysia uproots forest to build graveyard*. Retrieved June 22, 2007 from Reuters India Web site: http://in.today.reuters.com/news/newsArticle.aspx?type = worldNews&storyID = 2007–05–25T082650Z_01_NOOTR_RTRJONC_0_India-299856–1.xml&archived = False.

Folger, J.P., Poole, M.S., & Stutman, R.K. (1997). *Working through conflict*. 3rd edition. New York: Longman.

Forest of Dean Crime and Disorder Reduction Partnership. (2005). *Forest of Dean Crime and Disorder Reduction Strategy 2005–2008*. Retrieved December 4, 2007 from Forest of Dean Partnership Web site: http://www.forestofdeanpartnership.org.uk/content.asp?nav = 54&parent_directory_id = 29.

Forestry Commission. (2007). *Wild Woods at Jaguar Lount Wood*. Retrieved July 26, 2007 from ForestryCommission Website: http://www.forestry.gov.uk/website/ wildwoods.nsf/ LUWebDocsByKey/ England DerbyshireTheNationalForestJaguarLountWood.

Forestry Commission Wales. (2007). *FC Wales launches forest watch scheme to fight crime*. News release no. 9814, July 17, 2007. Retrieved December 4, 2007 from Forestry Commission

Web site: http://www.forestry.gov.uk/NewsRele.nsf/WebPressReleases/ BF456A115EF8FEA 180257313003D7A5C.

Foroohar, R. (2006). Unlikely boomtowns. *Newsweek* 2006(July 3–10), 50–62.

Forn Siðr. (2007). Retrieved July 10, 2006 from Forn Siðr Web site: http://www.fornsidr.dk/ index_ uk.html.

Forrest, M., & Konijnendijk, C.C. (2005). A history of urban forests and trees in Europe. In: C.C. Konijnendijk, K. Nilsson, T.B. Randrup, & J. Schipperijn (Eds.), *Urban forests and trees – a reference book* (pp. 23–48). Berlin: Springer.

Friedman, Th.L. (2000). *The lexus and the olive tree*. Updated and expanded edition. New York: Random House.

Gadet, J. (1992). Groen en de recreatieve functie van de openbare ruimte. *Groen* 1992(1), 9–13 (in Dutch).

Gallagher, W. (1993). *The power of place: How our surroundings shape our thoughts, emotions and actions*. New York: Harper Collins.

Gallis, C. (Ed.). (2005). *Proceedings, 1st European Cost E39 Conference: Forests, Trees, Human Health and Wellbeing*. Thessaloniki: Siokis.

Garside, P.L. (2006). Politics, ideology and the issue of open space in London, 1939–2000. In: P. Clark (Ed.), *The European city and green space. London, Stockholm, Helsinki and St Petersburg, 1850–2000* (pp. 68–98). Hants: Ashgate Historical Urban Studies.

Gathright, J., Yamada, Y., & Morita, M. (2007). Recreational tree-climbing programs in a rural Japanese community forest: social impacts and 'fun factors'. *Urban Forestry & Urban Greening* 6(3), 169–173.

Gazet van Antwerpen. (2007). *Marc Dutroux – Monster of zielepoot? Dossier Dutroux*. Retrieved October 29, 2007 from Gazet van Antwerpen Web site: http://www.gva.be/dossiers/d/dutroux/ dossier.asp (in Dutch).

Geen psychologisch gedoe. Het gewone vakantiekamp-recept werkt ook voor kinderen van vluchtelingen. (2006, August 11). *De Volkskrant* (in Dutch).

Geerts, P. (2002, November 16). Een schone slaapster. *De Financieel-Economische Tijd* (in Dutch).

Gehl, J. (2007). Public spaces for a changing public life. In: C.W. Thompson, & P. Travlou (Eds.), *Open space: People space. Engaging with the environment. Conference Proceedings* (pp. 3–9). London & New York: Taylor & Francis.

Gehrke, M. (2001). Der Park als Veranstaltungsort: Über Veranstaltungen im Grünen und. deren Verträglichkeit. *Stadt und Grün* 50(5), 325–332 (in German).

Gerritsen, F., Timmermans, W., & Visschedijk, P.A.M. (Eds.). (2004). *Groene Metropolen: het derde jaar, 2003* (pp. 64–65). Wageningen: Alterra.

Gesler, W. (2005). Therapeutic landscapes: an evolving theme. *Health & Place* 11(4), 295–297.

Gifford, R. (2002). *Environmental psychology: Principles and practice*. 3rd edition. Colville: Optimal Environments.

Gijsel, K. (2006). Openstelling van bos en natuur. Veelzeggende cijfers uit Nederland. *Bosrevue* 17(2006), 14–17 (in Dutch).

Gilbert, R. (2006). *A boom with a view. In: East London Green Grid: Essays* (pp. 12–13). London: Mayor of London.

Gini, R., & Selleri, B. (2007). *Management of urban forestry plantations in a climate change scenario: The experience of Parco Nord – Milan*. Paper presented at the conference Climate Change and Urban Forestry, Rome, December 15–16, 2007.

Graves, H.M., Watkins, R., Westbury, P., & Littlefair, P. (2001). *Cooling buildings in London: Overcoming the heat island*. London: CRC Ltd.

Green, G. (1996). *Epping forest through the ages*. 5th edition. Ilford, Essex: published by author.

Green Places. (2006). News item on Chigwell Row Wood Local Nature Reserve. *Green Places* 35 (2006), 6.

Grimbergen, C., Huibers, R., & Van der Peijl, D. (1983). *Amelisweerd: de weg van de meeste weerstand*. Rotterdam: Uitgeverij Ordeman (in Dutch).

Groene Gordel Front. (2006). *Lappersfort Museum*. Retrieved July 21, 2006 from Groene Gordel Front Web site: http://ggf.regiobrugge.be (in Dutch).

Groenewegen, P.P., Van den Berg, A.E., de Vries, S., & Verheij, R.A. (2006). Vitamin G: effects of green space on health, well-being, and social safety. *BMC Public Health* 2006(6), 49.
Goudsblom, J. (1992). *Vuur en beschaving*. Amsterdam: Meulenhoff (in Dutch).
Guldager, S. (2007). Fremtidens kirkegård – andet og mere end den kirkegård vi kender? *Landskab* 8(7), 156–157 (in Danish).
Guldemond, J.L. (1991). De Haarlemmerhout II: Beheersmogelijkheden in een oud stadsbos. *Nederlands Bosbouwtijdschrift* 63(4), 134–143 (in Dutch).
Gunnarsson, A., & Palenius, L. (2004). Stadsskog blir skolskog. *Gröna Fakta* 2004(3), II–IV (in Swedish).
Gustavsson, R. (2002). Afforestation in and near urban areas. In: T.B. Randrup, C.C. Konijnendijk, T. Christophersen, & K. Nilsson (Eds.), *COST Action E12: Urban Forests and Trees – Proceedings No 1* (pp. 286–314). Luxembourg: Office for Official Publications of the European Communities.
Gustavsson, R., Hermy, M., Konijnendijk, C., & Steidle-Schwahn, A. (2005). Management of urban woodland and parks – Searching for creative and sustainable concepts. In: C.C. Konijnendijk, K. Nilsson, T.B. Randrup, & J. Schipperijn (Eds.), *Urban forests and trees – a reference book* (pp. 369–397). Berlin: Springer.
Gustavsson, R., Mellqvist, H., & Åkerlund, U. (2004). *The Ronneby Brunn study. Testing a social-cultural based management planning, and the connoiseurs and action oriented approach as communicative tools in practice*. Alnarp: Department of Landscape Planning Alnarp, SLU.
Hagvaag, E. (2001, February 10). Ulve skaber ravage i Norge. *Politiken* (in Danish).
Hammer, J. (2002). First person global. *Newsweek* 2002(March 4), 5.
Hammitt, W.E. (2002). Urban forests and parks as privacy refuges. *Journal of Arboriculture* 28(1), 19–26.
Handlin, O. (1963). The modern city as a field of historical study. In: O. Handlin, & J. Burchard (Eds.), *The historian and the city* (pp. 1–26). Cambridge/London: The MIT Press.
Hands, D.E., & Brown, R.D. (2002). Enhancing visual preference of ecological rehabilitation sites. *Landscape and Urban Planning* 58(1), 57–70.
Hansen, K.B. (2005). *Fra cykelbarometer til tarzanjungle – et idékatalog om fysiske rammer, der fremmer bevægelse*. Copenhagen: Center for Forebyggelse, Sundhedsstyrelsen (in Danish).
Harrison, R.P. (1992). *Forests: The shadow of civilization*. Chicago & London: The University of Chicago Press.
Harrison, R.P. (2002). Hic Jacet. In: W.J.T. Mitchell (Ed.), *Landscape and power* (pp. 349–364). 2nd edition. Chicago & London: University of Chicago Press.
Hartig, T. (2004). Toward understanding the restorative environment as a health resource. In: *Open space: People space. Engaging with the environment. Conference proceedings*. OPENspace Research Centre, Edinburgh. Retrieved October 29, 2007 from OPENspace Web site: http://www.openspace.eca.ac.uk/conference/proceedings/summary/Hartig.htm.
Heitmann, B.L. (2000). Ten-year trends in overweight and obesity among Danish men and women aged 30–60 years. *International Journal of Obesity* 24(10), 1347–1352.
Hellinga, G. (1959). Het Nederlandse bos bedreigd! *Nederlands Bosbouwtijdschrift* 31(1). Reprinted in: *Nederlands Bosbouwtijdschrift* 75(2003), 4–9 (in Dutch).
Hellström, E., & Reunala, A. (1995). *Forestry conflicts from the 1950s till 1983*. EFI Research Report. Joensuu: European Forest Institute.
Hendriksen, I.J.M., Van Middelkoop, M., & Bervaes, J.C.A.M. (2003). *Wandelen tijdens de lunch*. TNO Rapport. Hoofddorp: TNO Arbeid (in Dutch).
Hennebo, D. (1979). *Entwicklung des Stadtgrüns von der Antike bis in die Zeit der Absolutismus*. Geschichte des Stadtgrüns, Band I. Hannover/Berlin: Patzer Verlag (in German).
Hennebo, D., & Hoffmann, A. (1963). *Geschichte der deutschen Gartenkunst*. Band III. Hamburg: Broschek Verlag (in German).
Hennebo, D., & Schmidt, E. (s.a.). *Entwicklung des Stadtgrüns in England von den frühen Volkswiesen bis zu den öffentlichen Parks im 19. Jahrhundert*. Geschichte des Stadtgrüns, Band III (1976–1979). Berlin: Patzer Verlag (in German).
Henwood, K., & Pidgeon, N. (2001). Talk about woods and trees: threat of urbanization, stability, and biodiversity. *Journal of Environmental Psychology* 21(2), 125–147.

Herzog, T.R. (1989). A cognitive analysis of preference for urban nature. *Journal of Environmental Psychology* 9, 27–43.
Heske, F. (1938). *German forestry*. New Haven: Yale University Press.
Heynen, N., Perkins, H.A., & Roy, P. (2006). The political ecology of uneven green space: the impact of political economy on race and ethnicity in producing environmental inequality in Milwaukee. *Urban Affairs Review* 42(1), 3–25.
Hibberd, B.G. (1989). *Urban forestry practice*. Forestry Commission Handbook No. 5. London: HMSO.
Hildebrand. (1998). *Camera Obscura*. Amsterdam: Athenaeum-Polak & Van Gennep (in Dutch).
Hiss, T. (1990). *The experience of place*. New York: Vintage Books/Random House.
Höppner, C., Frick, J., & Buchecker, M. (2007). Assessing psycho-social effects of participatory landscape planning. *Landscape and Urban Planning* 82(2–3), 196–207.
Hörnsten, L. (2000). *Outdoor recreation in Swedish forests – Implications for society and forestry*. Acta Universitatis Agriculturae Suecia, Silvestria 169. Uppsala: Swedish University of Agricultural Sciences.
Holm, S. (2000). *Anvendelse og betydning af byens parker og grønne områder*. Forest & Landscape Research 28(2000). Hørsholm: Skov & Landskab (in Danish).
Holscher, C.E. (1973). City forests of Europe. *Natural History* 82(11), 52–54.
Hosmer, R.S. (1988). Impressions of European forestry. Reprint of 1922 original. In: L. Fortmann, & J.W. Brice (Eds.), *Whose trees? Proprietary dimensions of forestry. Rural studies series of the Rural Sociological Society* (pp. 117–123). Boulder: Westview Press.
Houle, M.C. (1987). *One city's wilderness: Portland' Forest Park*. Jack Murdock Publication Series on the History of Science and Exploration in the Pacific Northwest. Portland: Oregon Historical Society Press.
Hunin, J. (2007, March 7). 'Ecologen maken ons stadje dood'. Activisten willen green ringweg door 'onbezoedeld' natuurgebeid. *De Volkskrant*, 4 (in Dutch).
HUR. (2005). *Forslag til Regionalplan 2005. Vision for Hovestadsregionen 2017 – en stærk og bæredytig region*. Valby: HUR (in Danish).
Husson, J.P. (1995). *Les forêts françaises*. Nancy: Press universitaires de Nancy (in French).
Ifversen, K.R.S. (2007, June 6). Byer med behov for en arkitekt. *Politiken* (in Danish).
INRA. (1979). *La forêt et la ville. Essai sur la forêt dans l'environnement urbain et industriel*. Versailles: Institut National de l'environnement urbain et industriel, Station de recherches sur la forêt et l'environnment (in French).
Irniger, M. (1991). *Der Sihlwald und sein Umland. Waldnutzung, Viehzucht und Ackerbau im Albisgebiet von 1400–1600*. Mitteilungen der Antiquarischen Gesellschaft in Zürich, Band 58. Zürich: Verlag Hans Rohr (in German).
Jaatinen, E. (1973). *Recreational utilization of Helsinki's forests*. Folia Forestalia 186. Helsinki: Finnish Forest Research Institute.
Jacobs, M. (2004). Metropolitan matterscape, powerscape and mindscape. In: G. Tress, B. Tress, B. Harms, P. Smeets, & A. van der Valk (Eds.), *Planning metropolitan landscapes – Concepts, demands, approaches* (pp. 26–37). Wageningen: Delta series, Volume 4.
Janse, G., & Konijnendijk, C.C. (2007). Communication between science, policy and citizens in public participation in urban forestry – Experiences from the NeighbourWoods project. *Urban Forestry & Urban Greening* 6(1), 23–40.
Jansen, I. (1995). Van Hollandse abstracties naar Russische werkelijkheid: 300 jaar tuin- en landschapsarchitectuur in Sint-Petersburg, deel 3. *Groen* 51(5), 26–32 (in Dutch).
Jansen-Verbeke, M. (Ed.). (2002). *De vrijetijdsfunctie 'van' en 'in' de stad*. Het 'Thuis in de Stad' project. Leuven: KU Leuven (in Dutch).
Jeanrenaud, S. (2001). *Communities and forest management in Western Europe*. A Regional Profile of WG – CIFM, the Working Group on Community Involvement in Forest Management (pp. 15–31). Gland: IUCN.
Jensen, F.J. (2003). *Friluftsliv I 592 skove og andre naturområder*. Skovbrugsserien nr. 32. Hørsholm: Skov & Landskab (in Danish).

Jensen, F.S. (1998). *Forest recreation in Denmark from the 1970s to the 1990s*. Frederiksberg & Hørsholm: The Royal Veterinary and Agricultural University & Danish Forest and Landscape Research Institute.

Jensen, F.S., Kaltenborn, B.P., & Sievänen, T. (1995). *Forest recreation research in Scandinavia*. Paper presented at 'Caring for the Forest: Research in a Changing World', IUFRO XX World Congress, August 6–12, 1995, Tampere, Finland.

Jókövi, E.M. (2000a). *Recreatie van Turken, Marokkanen en Surinamers in Rotterdam en Amsterdam: een verkenning van het vrijetijdsgedrag en van de effecten van de ethnische cultuur op de vrijetijdsbesteding*. Alterra-rapport 003. Wageningen: Alterra (in Dutch).

Jókövi, E.M. (2000b). *Vrijetijdsbesteding van allochtonen en autochtonen in de openbare ruimte. Een onderzoek naar de relatie met sociaal-economische en etnisch-culturele kenmerken*. Alterra-rapport 295. Wageningen: Alterra (in Dutch).

Jókövi, E.M., Bervaes, J., & Böttcher, S. (2002). *Recreatief gebruik van groene bedrijventerreinen. Een onderzoek onder werknemers en omwonenden*. Alterra-rapport 518. Wageningen: Alterra (in Dutch).

Joly, D. (2007). Gewestgrensoverschrijdend structuurplan Zoniënwoud laat op zich wachten. *Randbelangen*, May 23, 2007. Retrieved December 3, 2007 from Randbelangen Web site: http://www.randbelangen.org/?p = 347#more-347 (in Dutch).

Jones, N., & Baines, C. (2007). UK post-industrial regeneration. Developing strategies for functional green infrastructure. In: *Abstracts, 'New forests after old industries', 10th European Forum on Urban Forestry, May 16–19, 2007, Gelsenkirchen* (pp. 16–17). Dortmund: Landesbetrieb Wald und Holz Nordrhein-Westfalen.

Jones, O., & Cloke, P. (2002). *Tree cultures – The place of trees and trees in their place*. Oxford & New York: Berg.

Jongheid, J. (2000). Flevohout: dynamisch voorbeeldbos als ontmoetingsplaats voor houtproducent en –consument. *Nederlands Bosbouwtijdschrift* 72, 140–143 (in Dutch).

Jorgensen, A., & Anthopoulou, A. (2007). Enjoyment and fear in urban woodlands – Does age make a difference? *Urban Forestry & Urban Greening* 6(4), 267–278.

Jorgensen, A., Hitchmough, J., & Dunnet, N. (2006). Woodland as a setting for housing-appreciation and fear and the contribution of residential satisfaction and place identity in Warrington New Town, UK. *Landscape and Urban Planning* 79(3–4), 273–287.

Kaae, B., & Madsen, L.M. (2003). *Holdninger og ønsker til Danmarks natur*. By- og Landsplanserien nr. 21. Hørsholm: Skov & Landskab (in Danish).

Kalaora, B. (1981). Les salons verts: parcours de la ville à la forêt. In: M. Anselme, J.-L. Parisis, M., Péraldi, Y. Rochi, & B. Kalaora (Eds.), *Tant qu'il y aura des arbres. Pratiques et politiques de nature 1870–1960* (pp. 85–109). Recherches No. 45 (September 1981).

Kaplan, R. (2007). Employees' reaction to nearby nature at their workplace: The wild and the tame. *Landscape and Urban Planning* 82(1–2), 17–24.

Kaplan, R., & Kaplan, S. (1989). *The experience of nature: A psychological perspective*. Cambridge: Cambridge University Press.

Kaplan, S., & Talbot, J.F. (1983). Psychological benefits of a wilderness experience. In: I. Altman, & J.F. Wohlwill (Eds.), *Human behavior and environment, advances in theory and research*. Volume 6, Behavior and natural environment (pp. 163–203). New York: Plenum Press.

Kardell, L. (1998). *Anteckningar om friluftslivet på Norra Djurgården 1975–1996*. Umeå: Institutionen för Skoglig Landskapsvård, SLU (in Swedish).

Keil, P., Kowallik, C., Kricke, R., Loos, G.H., & Schlüpmann, M. (2007). Species diversity on urban-industrial brownfields with urban forest sectors compared with semi-natural habitats in wester Ruhrgebiet (Germany) – First results of investigations in flowering plants and various animal groups. In: *Abstracts, 'New forests after old industries', 10th European Forum on Urban Forestry, May 16–19, 2007, Gelsenkirchen* (pp. 31–32). Dortmund: Landesbetrieb Wald und Holz Nordrhein-Westfalen.

Kellner, U. (2000). Parks in Flandern. Historische Anlagen als Teil öffentlichens Grüns. *Stadt und Grün* 2000(3), 192–198 (in German).

Kennedy, J.J., Dombeck, M.P., & Koch, N.E. (1998). Values, beliefs and management of public forests in the Western world at the close of the twentieth century. *Unasylva* 49(192), 16–26.

Kennedy, M. (2005, April 19). Sculpture to inspire lazy Hampstead writers. *The Guardian*. Retrieved November 15, 2007 from The Guardian Web site: http://arts.guardian.co.uk/news/story/0,,1462939,00.html.

Khurana, N. (2006). Is there a role for trees in crime prevention? *Arborist News* August 2006, 26–28.

Kilvington, M., & Wilkinson, R. (1999). *Community attitudes to vegetation in the urban environment: A Christchurch case study*. Landscape Research Science Series no. 22. Lincoln: Manaaki Whenua Press.

Kirchengast, S., & Schober, E. (2006). Obesity among female adolescents in Vienna, Austria – the impact of childhood weight status and ethnicity. *BJOG* 113(10), 1188–1194.

Kirkebæk, M. (2002). The Danish afforestation policy and its results. In: T.B. Randrup, C.C. Konijnendijk, T. Christophersen, & K. Nilsson (Eds.), *COST Action E12: Urban Forests and Trees – Proceedings No 1* (pp. 277–285). Luxembourg: Office for Official Publications of the European Communities.

Kissane, M.J. (1998). Seeing the forest for the trees: land reclamation in Iceland. *Scandinavian Review*, spring 1998. Retrieved December 4, 2007 from FindArticles.com Web site: http://findarticles.com/p/articles/mi_qa3760/is_199804/ai_n8798492.

Kitaev, A. (2006). Red parks: green space in Leningrad, 1917–1990. In: P. Clark (Ed.), *The European city and green space. London, Stockholm, Helsinki and St Petersburg, 1850–2000* (pp. 289–303). Hants: Ashgate Historical Urban Studies.

Knecht, C. (2004). Urban nature and well-being: some empirical support and design implications. *Berkeley Planning Journal* 17, 82–108.

Koch, N. (1997). Forest, quality of life and livelihoods. In: *Social dimensions of forestry's contribution to sustainable development. Part F, Volume 5, Proceedings of the XI World Forestry Congress, October 13–22, 1997, Antalya* (pp. 27–32). Rome: FAO.

Kockelkoren, P. (1997). Bos-iconen. *Nederlands Bosbouwtijdschrift* 69, 247–261 (in Dutch).

Konijnendijk, C.C. (1994). Een korte geschiedenis van de Nederlandse stadsbossen. *Groen* 1994(10), 15–20 (in Dutch).

Konijnendijk, C.C. (1996). Van Fontainebleau tot Amelisweerd: de lessen van het stadsbos. *Nederlands Bosbouwtijdschrift* 68(2), 86–91 (in Dutch).

Konijnendijk, C.C. (1997). *Urban forestry: Overview and analysis of European urban forest policies: Part 1: conceptual framework and European urban forestry history*. EFI Working Paper 12. Joensuu: European Forest Institute.

Konijnendijk, C.C. (1998). Puzzelen in het Engelse landschap: het succesverhaal van de Community Forests. *De Boskrant* 28(4), 90–95 (in Dutch).

Konijnendijk, C.C. (1999). *Urban forestry in Europe: A comparative study of concepts, policies and planning for forest conservation, management and development in and around major European cities*. Doctoral dissertation. Research Notes No. 90. Joensuu: Faculty of Forestry, University of Joensuu.

Konijnendijk, C.C. (2000). Adapting forestry to urban demands – Role of communication in urban forestry in Europe. *Landscape and Urban Planning* 52(2–3), 89–100.

Konijnendijk, C.C. (2003). A decade of urban forestry in Europe. *Forest Policy and Economics* 5(3), 173–186.

Konijnendijk, C.C., & Flemish Forest Organisation (Eds.). (2001). *Communicating and financing urban forests. Proceedings of the 2nd and 3rd IUFRO European Forum on Urban Forestry, Aarhus (May 1999) and Budapest (May 2000)*. Brussels: Ministerie van de Vlaamse Gemeenschap, Afdeling Bos en Groen.

Konijnendijk, C.C., Ricard, R.M., Kenney, A., & Randrup, T.B. (2006). Defining urban forestry – A comparative perspective of North America and Europe. *Urban Forestry & Urban Greening* 4(3–4), 93–103.

Konijnendijk, C.C., Thorsen, B.J., Tyrväinen, L., Anthon, S., Nik, A.R., Haron, N., & Ghani, A.N. (2007). The right forest for the right city: decision-support for land-use planning through assessment of multiple forest benefits. *Allgemeine Forst und Jagdzeitung* 178(4), 74–84.

Koole, S.L., & Van den Berg, A.E. (2004). Paradise lost and reclaimed: an existential motives analysis of human-nature relations. In: J. Greenberg, S.L. Koole, & T. Pyszczynski (Eds.), *Handbook of experimental existential psychology* (pp. 86–103). New York: Guilford, New York.

Korpela, K., & Hartig, T. (1996). Restorative qualities of favorite places. *Journal of Environmental Psychology* 16, 221–233.

Korthals, M. (1998). Het alledaagse van exotische experimenten: vooronderstellingen en implicaties. *Nederlands Bosbouwtijdschrift* 70, 187–193 (in Dutch).

Kostof, S. (1999). *The city shaped. Urban patterns and meanings through history*. First paperback edition. London: Thames and Hudson.

Kotkin, J. (2005). *The city: A global history*. London: Weidenfeld & Nicolson.

Kotkin, J. (2006). Building up the Burbs. *Newsweek* 2006(July 3–10), 80–81.

Kowarik, I. (2005). Wild urban woodlands: towards a conceptual framework. In: I. Kowarik, & S. Körner (Eds.), *Wild urban woodlands – New perspectives for urban forestry* (pp. 1–32). Berlin: Springer.

Kowarik, I., & Körner, S. (Eds.). (2005). *Wild urban woodlands – New perspectives for urban forestry*. Berlin: Springer.

Krott, M. (1998). Urban forestry: management within the focus of people and trees. In: M. Krott, & K. Nilsson (Eds.), *Multiple-use of town forests in international comparison. Proceedings of the first European Forum on Urban Forestry, 5–7 May 1998, Wuppertal* (pp. 9–19). Wuppertal: IUFRO Working Group S.6.14.00.

Krott, M. (2004). Task-oriented comprehensive urban forestry – a strategy for forestry institutions. In: C.C. Konijnendijk, J. Schipperijn, & K.K. Hoyer (Eds.), *Forestry serving urbanized societies. Selected papers from the conference held in Copenhagen, Denmark, August 27–30, 2002* (pp. 79–90). IUFRO World Series Volume 14. Vienna: IUFRO.

Krott, M. (2005). *Forest policy analysis*. Dordrecht: Springer.

Kuo, F.E., Babaicoa, M., & Sullivan, W.C. (1998). Transforming inner-city landscapes. Trees, sense of place and preference. *Environment & Behavior* 30(1), 28–59.

Kupka, I. (2005). Silvicultural strategies in urban and periurban forests. Abstract. In: P. Neuhöfervá (Ed.), *Management of urban forests around large cities. Proceedings of abstracts, Prague, October 3–5, 2005* (p. 13). Prague: Faculty of Forestry and Environment, Czech University of Agriculture.

Kuyk, G.A. (1914). *De geschiedenis van het landgoed Sonsbeek bij Arnhem*. Reprint from: Bijdragen en Mededeelingen der Vereeniging 'Gelre', deel XVII (in Dutch).

Lafortezza, R., Corry, R.C., Sanesi, G., & Brown, R. (in press). Visual preferences and ecological assessments for designed alternative brownfield rehabilitations. Journal of Environmental Management. doi:10.1016/j.jenvman.2007.01.063.

Lambregts, L., & Wiersum, K.F. (2002). Community forestry in Nederland. *Nederlands Bosbouwtijdschrift* 74, 12–16 (in Dutch).

Landesbetrieb Wald und Holz NRW. (2006). *Project 'Industrial Forests Ruhr Region'. A model – The project – Presentation of project areas*. Project leaflet. Dortmund: LWH-NRW.

Lange, E., & Schaeffer, P.V. (2001). A comment on the market value of a room with a view. *Landscape and Urban Planning* 55(2001), 113–120.

Langers, F., de Boer, T.A., & Buijs, A.E. (2005). *Donkere nachten: de beleving van de nachtelijke duisternis door burgers*. Alterra Report 1137. Wageningen: Alterra (in Dutch).

Lappersfortbos vandaag of morgen ontruimd. (2002, September 17). *De Financieel-Economische Tijd*, p. 4 (in Dutch).

Lassøe, J., & Iversen, T.L. (2003). *Naturen i et hverdagslivsperspektiv. En kvantitativ interviewundersøgelse af forskellige danskeres forhold til naturen*. Faglig rapport fra DMU 437. Danmarks Miljøundersøgelser, Roskilde. Retrieved August 1, 2007 from DMU Web site: http://www2.dmu.dk/1_viden/2_Publikationer/3_fagrapporter/rapporter/FR437.pdf (in Danish).

Lawrence, H.W. (1993). The Neoclassical origins of modern urban forests. *Forest Conservation and History* 37, 26–36.

Lawrence, H.W. (2006). *City trees: A historical geography from the Renaissance through the nineteenth century*. Charlottesville & London: University of Virginia Press.

Layton, R.L. (1985). Recreation, management and landscape in Epping Forest: c. 1800–1984. *Field Studies* 6, 269–290.

Ledene, L. (2007). Het 'tweede grote Kom Over de Brug' gebeuren: verplanten van een symboolbos. *De Boskrant* 37(4), 14–16 (in Dutch).

Lee, K, & Pape, J. (2006). *Access to green areas affects health and well-being – Copenhagen on the Green Move*. Presentation at conference Urban Forestry for Human Health and Wellbeing, Copenhagen, June 27–30, 2006. Retrieved August 9, 2007 from University of Copenhagen Web site: http://en.sl.life.ku.dk/upload/kirsten_lee_002.pdf.

Lehmann, A. (1999). *Von Menschen und Bäumen. Die Deutschen und ihr Wald*. Reinbek: Rowohtl Verlag (in German).

Lento, K. (2006). The role of nature in the city: green space in Helsinki, 1917–60. In: P. Clark (Ed.), *The European city and green space. London, Stockholm, Helsinki and St Petersburg, 1850–2000* (pp. 188–206). Hants: Ashgate Historical Urban Studies.

Lepofsky, J., & Fraser, J.C. (2003). Building community citizens: claiming the right to place-making in the city. *Urban Studies* 40(1), 127–142.

Levent, T.B., & Nijkamp, P. (2004). *Urban green space policies: Performance and success conditions in European cities*. Preliminary version. Paper for the 44th European Congress of the European Regional Science Association, Regions and Fiscal Federalism, August 25–29, 2004, Porto.

Lieckfeld, C.-P. (2006). *Tatort Wald. Von einem, der auszog den Forst zu retten*. Frankfurt am Main: Westend Verlag (in German).

Lindsay, A., & House, S. (2005). *The tree collector. The life an explorations of David Douglas*. First paperback edition. London: Aurum Press.

Lipsanen, N. (2006). The seasonality of green space: the case of Uutela, Helsinki, c. 2000. In: P. Clark (Ed.), *The European city and green space. London, Stockholm, Helsinki and St Petersburg, 1850–2000* (pp. 229–246). Hants: Ashgate Historical Urban Studies.

Lloret, F., & Mari, G. (2001). A comparison of the medieval and the current fire regimes in managed pine forests of Catalonia (NE Spain). *Forest Ecology and Management* 141(2001), 155–163.

Löfström, I., Hamberg, L., Mikkola, N., & Koljonen, K. (2006). *Public participation, biodiversity and recreational values in urban forest planning in Finland*. Paper presented at IUFRO conference, Patterns and Processes in Forest Landscapes, Consequences of Human Management, Locorotondo, Italy, September 26–29, 2006.

Lörzing, H. (1992). *Van Bosplan to Floriade*. Rotterdam: Uitgeverij 010 (in Dutch).

Lohrberg, F. (2007). Landscape laboratory and biomass production – a 'Platform Urban Forestry Ruhrgebiet' demonstration project. In: *Abstracts, 'New forests after old industries', 10th European Forum on Urban Forestry, May 16–19, 2007, Gelsenkirchen* (pp. 20–21). Dortmund: Landesbetrieb Wald und Holz Nordrhein-Westfalen.

Longhurst, J. (2007). Chain reaction. *Green Places* 2007(36), 20–22.

Lorusso, L., Lafortezza, R., Tarasco, E., Sanesi, G., & Triggiani, O. (2007). Tipologie strutturali e caratteristiche funzionale delle aree verdi periurbane: il caso di studio della città di Bari. *L'Italia Forestale e Montane* LXII(4): 249–265 (in Italian, with English abstract).

Louv, R. (2006). *Last child in the woods: Saving our children from nature-deficit disorder*. New York: Algonquin Books.

Lub, J. (2000). Het Edese Bos: over vadertje Cats en frustraties in de A-lokaties. *Nederlands Bosbouwtijdschrift* 72, 6–10 (in Dutch).

Lusher, A. (2005, October 16). Spore wars: on the trail of the mushroom poachers. *The Sunday Telegraph*, 17.

Luttik, J. (2000). The value of trees, water and open space as reflected by house prices in The Netherlands. *Landscape and Urban Planning* 48(3/4), 161–167.

Maas J., Verheij R.A., Groenewegen P.P., De Vries S., & Spreeuwenberg P. (2006). Green space urbanity, and health: how strong is the relation? *Journal of Epidemiology and Community Health* 60, 587–592.

Maasland, F., & Reisinger, M. (2007). Pas op, grote grazers: over de verantwoordelijkheid van eigenaar, beheerder èn publiek. *Vakblad Natuur Bos Landschap* 4(1), 10–12 (in Dutch).
Macpherson, H. (2004). *Participation, practitioners and power: Community participation in North East Community Forests*. ESRC/ODPM Postgraduate Research Programme, Working Paper 13. London: ESRC/ODPM.
Macnaghten, P., & Urry, J. (1998). *Contested natures*. London: Sage.
Macnaghten, P., & Urry, J. (2000). Bodies in the Woods. *Body & Society* 6(3–4), 166–182.
Madas, A. (1984). The service functions. In: F.C. Hummel (Ed.), *Forest policy: A contribution to resource development* (pp. 127–159). The Hague: Nijhoff Junk/Publishers.
Mäkinen, K. (in prep.). Teenage experiences of public green areas in suburban Helsinki. Submitted to *Urban Forestry & Urban Greening* (in revision).
Mass hysteria after snake kill. (2006, September 7). Retrieved November 17, 2007 from News24.com Web site: http://www.news24.com/News24/World/News/0,,2–10–1462_1995155,00.html.
McConnachie, M., Shackleton, C., & McGregor, G.K. (2008). The extent of public green space and alien plant species in ten small towns of the Sub-Tropical Thicket Biome, South Africa. *Urban Forestry & Urban Greening* 7(1).
McIntyre, L. (2006). On the side of the angels. Landscape architects restore the Olmsted Woods at the Washington National Cathedral. *Landscape Architecture* 2006(7), 66–77.
Meier, U. (2007). Die rolle des Energieholzes in der Waldpolitik beider Basel (Essay). *Schweizerische Zeitschrift für Forstwesen* 158(7), 201–205 (in German, with English abstract).
Meierjürgen, U. (1995). *Freiraumerholung in Berlin*. Arbeitsmaterialen der Berliner Forsten 5. Berlin: Senatsverwaltung für Stadtentwicklung, Umweltschutz und Technologie (in German).
Meikar, T., & Sander, H. (2000). Die Forstwirtschaft in der Stadt Tallinn/Reval – ein historischer zugang. *Allgemeine Forst- und Jagdzeitung* 117(7), 124–131 (in German).
Merchant, C. (1996). Reinventing Eden: western culture as a recovery narrative. In: W. Cronon (Ed.), *Uncommon ground. Rethinking the human place in nature* (pp. 132–159). New York & London: W.W. Norton & Company.
Merriam-Webster Online Dictionary. (2007). *Entry for "city"*. Retrieved December 8, 2007 from Merriam-Webster Web site: http://www.m-w.com/dictionary/city.
Metz, T. (1998). *Nieuwe natuur. Reportages over veranderend landschap*. 2nd edition. Amsterdam: Ambo (in Dutch).
Mickenautisch, S., Legal, S.C., Yengopul, V., Bezerra, A.C., & Cruvinel, V. (2007). Sugar-free chewing gum and dental caries – A systematic review. *Journal of Applied Oral Science* 15(2), 83–88.
Milieudefensie. (2007). *'Bulderbos'*. Retrieved August 3, 2007 from Milieudefensie Web site: http://www.milieudefensie.nl/verkeer/doemee/bulderbos/ (in Dutch).
Milligan, C., & Bingley, A. (2007). Restorative places or scary spaces? The impact of woodland on the mental well-being of young adults. *Health & Place* 13, 799–811.
Millionærer og børns paradis. (2006, August 23). *Politiken* (in Danish).
Mind. (2007). *Ecotherapy: The green agenda for mental health*. London: Mind. Retrieved August 5, 2007 from Mind Web site: Publications.http://www.mind.org.uk/NR/rdonlyres/D9A930D2–30D4–4E5B-BE79–1D401B804165/0/ecotherapy.pdf.
Ministerium für Ländlichen Raum Baden-Württemberg. (1990). *Waldland Baden-Württemberg: Ein Überblick über den Wald und die Forstwirtschaft in Baden-Württemberg*. Stuttgart: Ministerium für Ländlichen Raum Baden-Württemberg (in German).
Mitchell, W.J.T. (Ed.). (2002). *Landscape and power. Space, place and landscape* (pp. vii–4). Chicago & London: University of Chicago Press.
Mitchell, R., & Popham, F. (2007). Greenspace, urbanity and health: relationships in England *Journal of Epidemiology and Community Health* 61, 681–683.
Moigneu, T. (2001). Junior Foresters Education Programme. In: C.C. Konijnendijk, & Flemish Forest Organisation (Eds.), *Communicating and financing urban forests. Proceedings of the 2nd and 3rd IUFRO European Forum on Urban Forestry, Aarhus (May 1999) and Budapest (May 2000)* (pp. 69–71). Brussels: Afdeling Bos & Groen, Ministerie van de Vlaamse Gemeenschap.

Møller, J. (1990). *Dyrehaven*. Copenhagen: Forlaget Cicero (in Danish).

Muir, R. (2005). *Ancient trees, living landscapes*. Stroud: Tempus.

Museumbos. (2007). Retrieved July 30, 2007 from Museumbos Web site http://www.museumbos.nl (in Dutch).

Myerscough, J. (1974). The recent history of the use of leisure time. In: I. Appleton (Ed.), *Leisure and research policy* (pp. 3–16). Edinburgh & London: Scottish Academic Press.

Nail, S. (2008). *Forest policies and social change. An English case study*. Berlin: Springer.

Nassauer, J.I. (1995). Messy ecosystems, orderly frames. *Landscape Journal* 14(2), 161–170.

Natuurbehoud. (2006). *Event announcement*. Issue autumn 2006. 's-Graveland: Natuurmonumenten (in Dutch).

Nederland Klimaatneutraal. (2007). *Hier.nu website*. Retrieved December 4, 2007 from Nederland Klimaatneutraal Web site: http://www.hier.nu (in Dutch).

Nehring, D. (1979). *Stadtparkanlagen in der erste Hälfte des 19. Jahrhunderts: ein Beitrag zur Kulturgeschichte des Landschaftsgarten*. Geschichte des Stadtgrüns. Band IV. Hannover & Berlin: Patzer Verlag (in German).

Neiss, Th. (2007). *Greeting*. In*: Abstracts, 'New forests after old industries', 10th European Forum on Urban Forestry, May 16–19, 2007, Gelsenkirchen* (pp. 5–7). Dortmund: Landesbetrieb Wald und Holz Nordrhein-Westfalen.

Newman, O. (1972). *Defensible space: Crime prevention through urban design*. New York: the Macmillan Company.

Nibbering, C., & Van Geel, R. (1993). Zucht naar de kathedraal: beleving en beheer in de Haarlemmerhout. *Groen* 1993(12), 26–29 (in Dutch).

Nielsen, A.B., & Jensen, R.B. (2007). Some visual aspects of planting design and silviculture across contemporary forest management paradigms – Perspectives for urban afforestation. *Urban Forestry & Urban Greening* 6(3), 143–158.

Nielsen, A.B., & Nilsson, K. (2007). Urban forestry for human health and wellbeing. Editorial. *Urban Forestry & Urban Greening* 6(4), 195–197.

Nielsen, T.S., & Hansen, K.B. (2007). Do green areas affect health? Results from a Danish survey on the use of green areas and health indicators. *Health & Place* 13, 839–850.

Niemi, M. (2006). Politicians, professionals and 'publics': conflicts over green space in Helsinki, c. 1950–2000. In: P. Clark (Ed.), *The European city and green space. London, Stockholm, Helsinki and St Petersburg, 1850–2000* (pp. 207–228). Hants: Ashgate Historical Urban Studies.

Nijen Twilhaar, H. (2007, September 24). Natuurbegraven wordt populair. *De Telegraaf*, p. T6 (in Dutch).

Nikolopoulou, M., Baker, S., & Steemers, K. (2001). Thermal comfort in outdoor urban spaces: understanding the human parameter. *Solar Energy* 70(3), 227–235.

Nilsson, K., Åkerlund, U., Konijnendijk, C.C., Alekseev, A., Guldager, S., Kuznetsov, E., Mezenko, A., & Selikhovkin, A. (2007a). Implementing green aid projects – the case of St Petersburg, Russia. *Urban Forestry & Urban Greening* 6(2), 93–101.

Nilsson, K., Baines, C., & Konijnendijk, C.C. (2007b). *Final report – COST Strategic Workshop, 'Health and the Natural Outdoors'*. Brussels: COST, European Science Foundation.

Nilsson, L. (2006a). Stockholm and green space 1850–2000: an introduction. In: P. Clark (Ed.), *The European city and green space. London, Stockholm, Helsinki and St Petersburg, 1850–2000* (pp. 99–110). Hants: Ashgate Historical Urban Studies.

Nilsson, L. (2006b). The Stockholm style: a model for the building of the city in the parks, 1930s–1960s. In: P. Clark (Ed.), *The European city and green space. London, Stockholm, Helsinki and St Petersburg, 1850–2000* (pp. 141–158). Hants: Ashgate Historical Urban Studies.

Nolan, B. (2006). *Phoenix Park. A history and guidebook*. Dublin: The Liffey Press.

Nolin, C. (2006). Stockholm's urban parks: meeting places and social contexts from 1860–1930. In: P. Clark (Ed.), *The European city and green space. London, Stockholm, Helsinki and St Petersburg, 1850–2000* (pp. 111–126). Hants: Ashgate Historical Urban Studies.

Northern Ireland Forest Service. (1987). *History of Belvoir Forest*. Retrieved July 31, 2007 from Forest Service Northern Ireland Web site: www.forestserviceni.gov.uk/history_of_belvoir.pdf.

Nowak, D.J. (2006). Institutionalizing urban forestry as a "biotechnology" to improve environmental quality. *Urban Forestry & Urban Greening* 5(2), 93–100.

Nowak, D.J., Walton, J.T., Dwyer, J.F., Kaya, L.G., & Myeong, S. (2005). The increasing influence of urban environments on US Forest management. *Journal of Forestry* 103(8), 377–382.

NUFU. (2005). *Trees Matter! Bringing lasting benefits to people in towns*. Wolverhampton: National Urban Forestry Unit.

O'Brien, L., & Murray, R. (2007). Forest School and its impacts on young children: case studies in Britain. *Urban Forestry and Urban Greening* 6(4), 249–265.

Olin, L. (2007). Foreword. In: C.W. Thompson, & P. Travlou (Eds.), *Open space: People space. Engaging with the environment. Conference Proceedings* (pp. xi–xvi). London & New York: Taylor & Francis.

Olwig, K.R. (1996). Reinventing common nature: Yosemite and Mount Rushmore – A meandering tale of a double nature. In: W. Cronon (Ed.), *Uncommon ground. Rethinking the human place in nature* (pp. 379–408). New York & London: W.W. Norton & Company.

Oosterbaan, A., Van Blitterswijk, H., & De Vries, S. (2005). *Gezond werk in het groen: Onderzoek naar de inzet van cliënten uit de zorg bij het beheer van bos, natuur en landschap*. Alterra rapport 1253. Wageningen: Alterra (in Dutch).

Opheim, T. (1984). Notes on the Oslomarka. In: O. Saastamoinen, S.-G. Hultman, N.E. Koch, & L. Matsson (Eds.), *Multiple use forestry in the Scandinavian countries* (pp. 39–43). Helsinki: Finnish Forest Research Institute.

Otto, H.J. (1998). Stadtnahe Wälder – kann die Forstwirtschaft sich noch verständlich machen? *Forst und Holz* 53(1), 19–22 (in German).

Oustrup, L. (2007). *Skovopfattelse blandt danskere og i skovlovgivningen*. Forest & Landscape Research no. 38. Hørsholm: Danish Centre for Forest, Landscape and Planning, University of Copenhagen (in Danish).

Paasman, J. (1997). De Natuurverkenning en het groene perspectief: Het Sportfondsenbos. *Nederlands Bosbouwtijdschrift* 69, 150–155 (in Dutch).

Paci, M. (2002). *L'uomo e la foresta*. Rome: Meltemi Editore (in Italian).

Paris, J.D. (1972). The citification of the forest. *Canadian Pulp and Paper Magazine* 9, 119–122.

Parque Amazónia, Belém. (2006). Entry in overview of winners of ASLA Awards 2005. *Landscape Architecture* 2006(7), 79.

Pemmer, H., & Lackner, N. (1974). *Der Prater: von den Anfängen bis zur Gegenwart*. 2nd edition, edited by Düriegel, G., Sackmauer, L. Wiener Heimatkunde. Vienna: Jugend & Volk (in German).

Pendleton, M.R., & Thompson, H.L. (2000). The criminal career of park and recreation hotspots. *Parks & Recreation* 35(7), 56–63.

Perlin, J. (1989). *A forest journey. The role of wood in the development of civilization*. Cambridge & London: Harvard University Press.

Perlis, A. (Ed.). (2006). Forests and human health. *Unasylva* 224.

Perlman, M. (1994). *The power of trees. The reforesting of the soul*. Woodstock: Spring Publications.

Pirnat, J. (2005). Multi-functionality in urban forestry – A dream or a task? In: C.C. Konijnendijk, J. Schipperijn, & K. Nilsson (Eds.), *COST Action E12: Urban forests and trees – Proceedings No 2* (pp. 101–118). Luxembourg: Office for Official Publications of the European Communities.

Politie ontruimt Lappersfortbos na immense machtsontplooiing. (2002, October 15). *Gazet van Antwerpen* (in Dutch).

Pollan, M. (1991). *Second nature: A gardener's education*. New York: Atlantic Monthly Press.

Ponting, C. (1991). *A green history of the world*. London: Penguin Books.

Porteous, A. (2002). *The forest in folklore and mythology*. Reprint of 1928 original. Mineola: Dover Publications.

Pouta, E., Neuvonen, M., & Sievänen, T. (2007). Prognosis and scenarios of outdoor recreation. In: *Abstracts, IUFRO conference Integrative Science for Integrative Management, Saariselkä, Finland, August 14–20, 2007*. Vantaa: Metla.

Preston, I. (2007). A Northern network. *Green Places* 2007(36), 16–18.

Preston, S. (2007). Wildlife on the doorstep. *Green Places* 2007(34), 36–38.

Priestley, G., Montenegro, M., & Izquierdo, S. (2004). Greenspace in Barcelona: An analysis of user preferences. In: B. Martens, & A.G. Keul (Eds.), *Evaluation in progress – Strategies for environmental research and implementation*. IAPS conference July 7–9, 2004. CD rom.

Prijs Lappersfortbos nog steeds struikelblok. (2002, October 15). *De Financieel-Economische Tijd*, p. 4 (in Dutch).

Proctor, J.D. (1996). Whose nature? The contested moral terrain of ancient forests. In: W. Cronon (Ed.), *Uncommon ground. Rethinking the human place in nature* (pp. 269–297). New York & London: W.W. Norton & Company.

Pröbstl, U. (2004). Trends in outdoor recreation – Should the planning for open spaces follow them? In: *Proceedings of Open Space: People Space, an international conference*. Retrieved December 1, 2006 from OPENspace Web site: http://www.openspace.eca.ac.uk/conference/proceedings/summary/Probstl.htm.

Pröbstl, U. (2007). Forests in balance? Forest under the spell of economic, ecological and recreational requirements – Considerations about the European Model. *Allgemeine Forst und Jagdzeitung* 178(4), 68–73 (in German).

Profous, G., & Rowntree, R. (1993). Structure and management of the urban forest in Prague. *Unasylva* 44(173), 33–38.

PROGRESS. (2007). *PROGRESS project information*. Retrieved July 31, 2007 from PROGRESS Web site: http://www.progress-eu.info/uk.htm.

Proshansky, H.H., Fabian, A.K., & Kaminoff, R. (1983). Place identity: physical world socialization of the self. *Journal of Environmental Psychology* 3(3), 57–83.

Raad voor het Landelijk Gebied. (2002). *Voor boeren, burgers en buitenlui. Advies over de betekenis van sociaal-culturele ontwikkelingen voor het landelijk gebied* (pp. 65–69). RLG 02/08. Den Haag: Raad voor het Landelijk Gebied (in Dutch).

Rackham, O. (2004). *Trees and woodland in the British landscape. The complete history of Britain's trees, woods & hedgerows*. Revised edition. New York: Phoenix Press.

Ramdharie, S. (1995, December 29). In Scheveningse Bosjes hangt 'een enge sfeer'. Bewoners zijn overlast door homobaan beu. *De Volkskrant* (in Dutch).

Randrup, T.B., Konijnendijk, C., Kaennel Dobbertin, M., & Prüller, R. (2005). The concept of urban forestry in Europe. In: C.C. Konijnendijk, K. Nilsson, T.B. Randrup, & J. Schipperijn (Eds.), *Urban forests and trees – a reference book* (pp. 9–20). Berlin: Springer.

Ransford, M. (1999). Fear spreading faster than Lyme disease. Newscenter, Ball State University, November 10, 1999. Retrieved December 3, 2007 from Ball State University Web site: http://www.bsu.edu/news/article/0,1370,-1019-974,00.html.

Rasmussen, K.R., & Hansen, K. (Eds.). (2003). *Grundvand fra skove – muligheder og problemer*. Skovbrugsserien. Hørsholm: Skov & Lanskab (in Danish).

Reeder, D.A. (2006a). London and green space. 1850–2000: an introduction. In: P. Clark (Ed.), *The European city and green space: London, Stockholm, Helsinki and St Petersburg, 1850–2000* (pp. 30–40). Hants: Ashgate Historical Urban Studies.

Reeder, D.A. (2006b). The social construction of green space in London prior to the Second World War. In: P. Clark (Ed.), *The European city and green space: London, Stockholm, Helsinki and St Petersburg, 1850–2000* (pp. 41–67). Hants: Ashgate Historical Urban Studies.

Région Wallonne (Division de la Nature et des Forêts), Vlaams Gewest, & Région Bruxelles Capitale. (2007). *Zoniënwoud*. Retrieved December 27, 2007 from Zonïenwoud Web site: http://www.sonianforest.be (in Dutch and French, with English introduction).

Relf, D. (1992). Human issues in horticulture. *HortTechnology* 2(2). Retrieved November 14, 2007 from Virginia Tech Web site: http://www.hort.vt.edu/human/hihart.htm.

Remembering Sarajevo's children. (1997). American Forests 103(3), 9.

Reneman, D.-D., Visser, M., Edelmann, E., & Mors, B. (1999). *Mensenwensen; De wensen van Nederlanders ten aanzien van natuur en groen in de leefomgeving*. Natuur als leefomgeving – Operatie Boomhut. Reeks Operatie Boomhut nr. 6. Hilversum & Den Haag: Intomart en Ministerie van LNV (in Dutch).

Röbbel, H. (1967). The role of shooting for recreation in Germany. In: *XIV. IUFRO Kongress, München 1967, Referate, VII, Section 26* (pp. 310–329). Munich: DVFFA. Sekretariat für de IUFRO-Kongress 1967.

Romeijn, W.F. (2005). Nieuwe sport in natuur- en bosterreinen: Geocoaching. *Vakblad Natuur-Bos-Landschap* 2(3), 2–5 (in Dutch).

Ronge, V. (1998). Urban areas and their town forests: the impact of values, interests and political dynamics on urban forestry. In: M. Krott, & K. Nilsson (Eds.), *Multiple use of town forests in international comparison. Proceedings of the first European Forum on Urban Forestry, May 5–7, 1998, Wuppertal* (pp. 31–41). Wuppertal: IUFRO Working Group S.6.14.00.

Rook, A. (2007). Productive landscapes. *Green Places* 2007(34), 9.

Roovers, P. (2005). *Impacts of outdoor recreation on ecosystems: Towards an integrated approach*. Doctoraalproefschrift nr. 650. Leuven: Faculteit Bio-ingenieurswetenschappen, K.U. Leuven.

Rotenberg, R. (1995). *Landscape and power in Vienna*. Baltimore: The Johns Hopkins University Press.

Ruffier-Reynie, C. (1992). Fontainebleau, merveille en grand péril. *Combat Nature* 99, 32–37 (in French).

Ruffier-Reynie, C. (1995). Un parc national pour la forêt de Fontainebleau? *Combat Nature* 111, 25–30 (in French).

Runte, A. (1987). *National Parks: The American experience*. 2nd, revised edition. Lincoln & London: University of Nebraska Press.

Rydberg, D. (1998). *Urban forestry in Sweden – Silvicultural aspects focusing on young forests*. Doctoral thesis. Acta Universitatis Agriculturae Sueciae – Silvestria 73. Umeå: Swedish University of Agricultural Sciences.

Rydberg, D., & Falck, D. (1998). Designing the urban forest of tomorrow: pre-commercial thinning adapted for use in urban areas in Sweden. *Arboricultural Journal* 22, 147–171.

Sanesi, G., Lafortezza, R., Marziliano, P.A., Ragazzi, & L. Mariani. (2007). Assessing the current status of urban forest resources in the context of Pardo Nord, Milan, Italy. *Landscape and Ecological Engineering* 2007(3), 187–198.

Sapochkin, M.S., Kiseleva, V.V., Syriamkina, O.V., & Nitikin, V.F. (2004). Mapping the intensity of recreation impact in the NP Losiny Ostrov, Moscow. In: T. Sievänen, J. Erkkonen, J. Jokimäki, J. Saarinen, S. Tuulentie, & E. Virtanen (Eds.), *Policies, methods and tools for visitor management. Proceedings of the Second International Conference in Monitoring and Management of Visitor Flows in Recreational and Protected Areas, June 16–20, Rovaniemi, Finland* (pp. 45–50). Working Papers of the Finnish Forest Research Institute. Vantaa: Metla.

Schama, S. (1995). *Landscape and memory*. London: HarperCollins Publishers.

Schantz, P. (2006). The formation of National Urban Parks: a Nordic contribution to sustainable development? In: P. Clark (Ed.), *The European city and green space: London, Stockholm, Helsinki and St Petersburg, 1850–2000* (pp. 159–174). Hants: Ashgate Historical Urban Studies.

Schantz, P., & Silvander, U. (Eds.). (2004). *Forskning och utbildning inom friluftsliv. Utredning och förslag*. Stockholm: FRISAM, Friluftsorganisationer i Samverkan (in Swedish).

Schriver, N.B. (2006). *The significance of the woods in the experience of pleasure, freedom, desire and courage as important in the rehabilitation process*. Paper presented at the conference Urban Forestry for Human Health and Wellbeing, Copenhagen, June 27–30, 2006. Retrieved August 9, 2007 from Danish Centre for Forest, Landscape and Planning Web site: http://en.sl.life.ku.dk/upload/n_schriver.pdf.

Schroeder, H.W. (1992). The spiritual aspect of nature: a perspective from depth psychology. In: *Proceedings of Northeastern Recreation Research Symposium, April 7–9, 1991, Saratoga Springs, NY* (pp. 25–30). Philadelphia: U.S. Department of Agriculture, Forest Service, Northeastern Forest Experiment Station.

Schroeder, H.W. (2001). *Mythical dimensions of trees. Their relation to the human psyche as reflected in dreams, myths and cultural traditions*. Retrieved August 16, 2006 from Garden State Environet Web site: http://www.gsenet.org/library/08for/treemyth.php.

Schulte, A.G., & Schulte-van Wersch, C.J.M. (2006). *Monumentaal groen. Kleine cultuurgeschiedenis van de Arnhemse parken*. 2nd edition. Reeks Arnhemse Monumenten 7. Arnhem: Uitgeverij Matrijs (in Dutch).

Seeland, K., Moser, K., Scheutle, H., & Kaiser, F.G. (2002). Public acceptance of restrictions imposed on recreational activities in the peri-urban Nature Reserve Sihlwald, Switzerland. *Urban Forestry & Urban Greening* 1(1), 49–57.

Semenov, K. (2006). St. Petersburg's parks and gardens, 1850–1917. In: P. Clark (Ed.), *The European city and green space. London, Stockholm, Helsinki and St. Petersburg, 1850–2000* (pp. 272–288). Hants: Ashgate Historical Urban Studies.

Shin, W.S. (2007). The influence of forest view through a window on job satisfaction and job stress. *The Scandinavian Journal of Forest Research* 22(3), 248–253.

Sieghardt, M., Mursch-Radlgruber, E., Paoletti, E., Couenberg, E., Dimitrakopoulus, A., Rego, F., Hatzistathis, A., & Randrup, T.B. (2005). The abiotic environment: impact of urban growing conditions on urban vegetation. In: C.C. Konijnendijk, K. Nilsson, T.B. Randrup, & J. Schipperijn (Eds.), *Urban Forests and Trees – A reference book* (pp. 313–315). Berlin: Springer.

Sievänen, T., Pouta, E., & Neuvonen, M. (2003). Regional similarities and differences in Finnish outdoor recreation behavior. *Terra* 115(4), 259–273.

Silfverberg, K., Möller, J., Nyhuus, S., Müllerström, J., Christensson, K., Nerpin, L., Gudmundsson, H., Siggurdsson, Ö., Nielsen, J.D., & Nielsen, P. (2003). *Nordiske byers miljøindikatorer. Nordisk Storbysamarbejde 2003 – et fællesprojekt mellem 7 storbyer I Norden: Göteborg, København, Oslo, Stockholm, Reykjavik, Malmö og Helsingfors*. Copenhagen: Københavns Kommune (in Danish).

Simon, B. (2001). Increasing the value of trees in the Forest of Belfast. In: K. Collins, & C.C. Konijnendijk (Eds.), *Planting the idea – the role of education in urban forestry. Proceedings of COST Action E12 'Urban Forests and Trees' seminar, Dublin, March 23, 2000* (pp. 12–20). Dublin: The Tree Council of Ireland.

Simon, B. (2005). Chapter 2 – The history of Belvoir Park. In: B. Simon (Ed.), *A treasured landscape – the heritage of Belvoir Park* (pp. 16–42). Belfast: The Forest of Belfast, Belfast.

Simson, A.J. (1997). The post-romantic landscape of Telford New Town. *Landscape and Urban Planning* 52(2–3), 189–197

Simson, A. (2001a). Art, a gateway that leads to understanding the urban forest? In: C.C. Konijnendijk, & Flemish Forest Organisation (Eds.), *Communicating and financing urban forests. Proceedings of the 2nd and 3rd IUFRO European Forum on Urban Forestry, Aarhus (May 1999) and Budapest (May 2000)* (pp. 23–32). Brussels: Afdeling Bos en Groen, Ministerie van de Vlaamse Gemeenschap.

Simson, A. (2001b). The ARBRE project, Yorkshire, United Kingdom. In: C.C. Konijnendijk, & Flemish Forest Organisation (Eds.), *Communicating and financing urban forests. Proceedings of the 2nd and 3rd IUFRO European Forum on Urban Forestry, Aarhus (May 1999) and Budapest (May 2000)* (pp. 199–200). Brussels: Afdeling Bos en Groen, Ministerie van de Vlaamse Gemeenschap.

Simson, A. (2005a). The White Rose Forest – A catalyst for the regeneration of the region. In: C.C. Konijnendijk, J. Schipperijn, & K. Nilsson (Eds.), *COST Action E12: Urban forests and trees – Proceedings No. 2* (pp. 237–248). Luxembourg: Office for Official Publications of the European Communities.

Simson, A. (2005b). Urban forestry in Europe: innovative solutions and future potential. In: C.C. Konijnendijk, K. Nilsson, T.B. Randrup, & J. Schipperijn (Eds.), *Urban forests and trees – a reference book* (pp. 479–520). Berlin: Springer.

Simson, A. (2007a). The importance of seeking out and responding to woodland heritage cues when developing new, viable urban forestry initiatives for post-industrial landscapes. In: *Abstracts, 'New forests after old industries', 10th European Forum on Urban Forestry, May*

16–19, 2007, Gelsenkirchen (pp. 22–23). Dortmund: Landesbetrieb Wald und Holz Nordrhein-Westfalen.
Simson, A. (2007b). Urban Wildscapes. Conference report. *Green Places* 2007(409), 38–39.
Skov-Petersen, H., & Jensen, F.S. (2007). An empirical study of recreational route choices. In: *Abstracts, IUFRO conference Integrative Science for Integrative Management, Saariselkä, Finland, August 14–20, 2007*. Vantaa: Metla.
Slabbers, S., Bosch, J.W., Van den Hamer, J.H., & Ulijn, J. (1993). *Grote bossen bij Europese steden: onderzoek naar de landschappelijke kwaliteit van de bossen Epping Forest, Grunewald and Forêt de Saint Germain*. Wageningen: IKCN (in Dutch).
Slater, C. (1996). Amazonia as Edenic Narrative. In: W. Cronon (Ed.), *Uncommon ground. Rethinking the human place in nature* (pp. 114–131). New York & London: W.W. Norton & Company.
Smith, A. (2007a). In search of liveable communities. Letter. *Green Places* 2007(35), 14.
Smith, S. (2007b). Making it safe. *Green Places* 2007(34), 47.
Sörensen, A.-B., & Wembling, M. (1996). Allergi och stadsgrönska. *Gröna Fakta* 3/96. Alnarp: Movium (in Swedish).
Sonsbeek krijgt ruime voldoende. (1996, February 10). *De Gelderlander* (in Dutch).
Spencer, C., & Woolley, H. (2000). Children and the city: a summary of recent environmental psychology research. Summary. *Child: Care, Health and Development* 26(3), 181.
Spiegal, B. (2007). The starting point for play. *Green Places* 2007(34), 26–29.
Spirn, A.W. (1996). Constructing nature: the legacy of Frederick Law Olmsted. In: W. Cronon (Ed.), *Uncommon ground. Rethinking the human place in nature* (pp. 91–113). New York & London: W.W. Norton & Company.
Spongberg, S. (1990). *A reunion of trees – The discovery of exotic plants and their introduction into North American and European landscapes*. Cambridge: Harvard University Press.
Stigsdotter, U., & Grahn, P. (2002). Landscape planning and stress. *Urban Forestry & Urban Greening* 2(1), 1–18.
Stisen, V. (2003, March 17). Motionister indtog afspærret ø. *Politiken*, 1–2 (in Danish).
Strzygowski, W. (1967). Die Wiederbewaldung der Küsten Griechenlands, eine Voraussetzung der Steigerung des Tourismus und damit der künftig wichtigsten Einkommensquelle. In: *XIV. IUFRO Kongress, München 1967, Referate, VII, Section 26* (pp. 362–381). Munich: DVFFA. Sekretariat für de IUFRO-Kongress 1967 (in German).
Suau, L., & Confer, J. (2005). Parks and the geography of fear. In: J.G. Peden, & R.M. Schuster (Eds.), *Proceedings of the 2005 Northeastern Recreation Research Symposium, April 10–12, Bolton Landing, NY* (pp. 273–278). GTR-NE-341. Newtown Square: U.S. Forest Service, Northeastern Research Station.
Sugiyama, T., & Thompson, C.W. (2007). Older people's health, outdoor activity and supportiveness of neighbourhood environments. *Landscape and Urban Planning* 82(2–3), 168–175.
Swarnasinghe, K.M.I. (2005). *World's oldest historical sacred bodhi tree at Anuradhapura*. Erewwala Pannipitiya: Chaga Publications.
Szramka, J. (1995). Forest management and nature conservation on the area of the Regional Directorate of State Forests in Gdansk. In: *Proceedings of the XIII conference of the Union of European Foresters, Gdansk, May 1995*. Warsaw: Warsaw Agricultural University.
Takeuchi, K., Brown, R.D., Washitani, A., Tsunekawa, A., & Yokohari, M. (Eds.). (2003). Satoyama, the traditional landscape of Japan. Berlin: Springer.
Talen, E. (1998). Sense of community and neighbourhood form: an assessment of the social doctrine of new urbanism. *Urban Studies* 36(8), 1361–1379.
Talking with the author Richard Louv. (2007). *The Forestry Source* 2007(June), 8–9.
Tam, J.N. (1980). Housing reform and the emergence of town planning in Britain before 1914. In: A. Sutcliffe (Ed.), *The rise of modern urban planning, 1800–1914* (pp. 71–97). London: Mansell.
Taplin, K. (1989). *Tongues in trees. Studies in literature and ecology*. Bideford: Green Book.
Taylor, A.F., & Kuo, F.E. (2006). Is contact with nature important for healthy child development? State of evidence. In: Ch. Spencer, & M. Blades (Eds.), *Children and their environments: Learning, using and designing spaces* (pp. 124–140). Cambridge: Cambridge University Press.
Teletekst. (1999, July 8). *News item, EO-Radio*. Hilversum: NOS/EO (in Dutch).

Terror detectives 'find bomb kit'. (2006, August 17). Retrieved August 8, 2007 from BBC News Web site: http://news.bbc.co.uk/2/hi/uk_news/5261086.stm.

'*The Battle of Epping* Forest'. (2007). Wikipedia entry. Retrieved July 24, 2007 from Wikipedia Web site: http://en.wikipedia.org/wiki/The_Battle_of_Epping_Forest.

The Corporation of London. (1993). *Official guide to Epping Forest*. London: Guildhall.

Theil, S. (2006). The new jungles. *Newsweek*, 2006(July 3–10), 75–77.

The SAUL Partnership. (2005). *Vital urban landscapes – the vital role of sustainable and accessible urban landscapes in European city regions*. Final report of the Sustainable & Accessible Urban Landscapes (SAUL) Partnership. London: Groundwork.

Thomas, D. (1990). The edge of the city. *Transactions of the Institute of British Geographers* 15(2), 131–138.

Thomas, K. (1984). *Man and the natural world: Changing attitudes in England from 1500 to 1800*. Harmondsworth: Penguin.

Thompson, C.W., Travlou, P., & Roe, J. (2006). *Free-range teenagers: The role of wild adventure space in young people's lives*. Final report, prepared for Natural England. Edinburgh: OPENspace.

Thompson, J.L., & Thompson, J.E. (2003). The urban jungle and allergy. *Immunology and Allergy Clinics of North America* 23(3), 371–387.

Travlou, P., & Thompson, C.W. (2007). Preface. In: C.W. Thompson, & P. Travlou (Eds.), *Open space: People space. Engaging with the environment. Conference Proceedings* (pp. xvii–xix). London & New York: Taylor & Francis.

Trébucq, M. (1995). Incidents. Sauver Fontainebleau no, 132, supplement of *Télérama* 2372, 2–5 (in French).

Treib, M. (2002). Place, time and city trees. In: T.B. Randrup, C.C. Konijnendijk, T. Christophersen, & K. Nilsson (Eds.), *COST Action E12: Urban Forests and Trees – Proceedings No. 1* (pp. 64–79). Luxembourg: Office for Official Publications of the European Communities.

Trodden, L. (2007). Integrating art. *Green Places*, 2007(37), 24–25.

Truhlár, J. (1997). *The Rícmanice Arboretum and Memorial of Trees*. Mendel Brno: University of Agriculture and Forestry.

Tuan, Y.-F. (2007). *Space and place. The perspective of experience*. 5th edition. Minneapolis & London: University of Minnesota Press.

Tyrväinen, L. (1999). *Monetary valuation of urban forest amenities in Finland*. Academic dissertation. Research Papers 739.Vantaa: Finnish Forest Research Institute.

Tyrväinen, L., Gustavsson, R., Konijnendijk, C.C., & Ode, Å. (2006). Visualization and landscape laboratories in planning, design and management of urban woodlands. *Forest Policy and Economics* 8(8), 811–823.

Tyrväinen, L., Mäkinen, K., & Schipperijn, J. (2007). Tools for mapping social values of urban woodlands and other green areas. *Landscape and Urban Planning* 79(1), 5–19.

Tyrväinen, L., Pauleit, S., Seeland, K., & De Vries, S. (2005). Benefits and uses of urban forests and trees. In: C.C. Konijnendijk, K. Nilsson, T.B. Randrup, & J. Schipperijn (Eds.), *Urban forests and trees – A reference book* (pp. 81–114). Berlin: Springer.

Tyrväinen, L., Silvennoinen, H., & Kolehmainen, O. (2003). Ecological and aesthetic values in urban forest management. Urban Forestry & Urban Greening 1(3), 135–149.

Ulrich, R.S. (1984). View through a window may influence recovery from surgery. *Science* 224 (4647), 420–421.

Umweltdachverband für das Biosphärenpark Wienerwald Management, & Österreichische UNESCO Kommission. (2006). *Leben im Biosphärenpark Wienerwald. Modellregion der Nachhaltigkeit. Vienna*. Retrieved October 29, 2007 from Biosphärenpark Wienerwald Web site: http://www.biosphaerenpark-wienerwald.org/cms/_data/S1–34.pdf (in German).

UN DESA Population Division. (2005). *World urbanization prospects: The 2005 revision*. Retrieved December 13, 2007 from United Nations Web site: /www.un.org/esa/population/publications/WUP2005/2005wup.htm.

UNESCO. (2001). *Universal declaration on cultural diversity*. Signed at Paris, November 2, 2001. Retrieved July 31, 2006 from UNESCO Web site: http://unesdoc.unesco.org/images/0012/001271/127160m.pdf.

Vallejo, R. (2005). Managing forest fires near urban areas in Mediterranean countries. In: C.C. Konijnendijk, J. Schipperijn, & K. Nilsson (Eds.), *COST Action E12: Urban forests and trees – Proceedings No. 2* (pp. 225–233). Luxembourg: Office for Official Publications of the European Communities.

van der Ben, D. (2000). *La forêt de Soignes: Passé, présent, avenir*. 2nd edition. Brussels: Éditions Racines (in French).

Van den Berg, A.E., & Ter Heijne, M. (2004). Angst voor de natuur: een theoretische en empirische verkenning. *Landschap* 2004(3), 137–145 (in Dutch).

Van den Berg, J. (2000). Speelbossen bij Staatsbosbeheer. *Nederlands Bosbouwtijdschrift* 72: 38–39 (in Dutch).

Van den Berg, R. (2005). Planning new forests in the Netherlands. In: C.C. Konijnendijk, J. Schipperijn, & K. Nilsson (Eds.), *COST Action E12: Urban forests and trees – Proceedings No. 2* (pp. 55–63). Luxembourg: Office for Official Publications of the European Communities.

Van den Ham, M.H.A. (1997). Beleidsvorming voor stadsbossen in Amsterdam en Arnhem: urban forestry. *Nederlands Bosbouwtijdschrift* 69(1), 29–37 (in Dutch).

Van der Plas, G. (Ed.). (1991). *De openbare ruimte van de stad*. Amsterdam: Stadsuitgeverij Amsterdam (in Dutch).

Van Herzele, A. (2005). *A tree on your doorstep, a forest in your mind. Greenspace planning at the interplay between discourse, physical conditions, and practice*. Doctoral dissertation. Wageningen: Chair of Communication and Innovation Studies, Wageningen University.

Van Herzele, A., Collins, K., & Tyrväinen, L. (2005). Involving people in urban forestry – A discussion of participatory practices throughout Europe. In: C.C. Konijnendijk, K. Nilsson, T.B. Randrup, & J. Schipperijn (Eds.), *Urban forests and trees – a reference book* (pp. 207–228). Berlin: Springer.

Van Herzele, A., & Wiedemann, T. (2003). A monitoring tool for the provision of accessible and attractive urban green spaces. *Landscape and Urban Planning* 63(2003), 109–126.

Van Kerckhove, B., & Zwaenepoel, J. (1994). *9 wandelingen in het Zoniënwoud*. Series: Op stap in Vlaams-Brabant. Leuven: Toeristische Federatie van Brabant (in Dutch).

Van Otterloo, R. (2006). 'Sonsbeek 'moet' elke drie jaar'. *Infobulletin Parken Sonsbeek, Zijpendaal en Gulden Bodem* 16(4), 3–4 (in Dutch).

Van Rooijen, M. (1990). *De wortels van het stedelijk groen: een studie naar het ontstaan van de Nederlandse Groene stad*. Utrecht: Vakgroep Stads- en Arbeidsstudie, Rijksuniversiteit Utrecht (in Dutch).

Van Winsum-Westra, M., & De Boer, T.A. (2004). *(On)veilig in bos & natuur. Een verkenning van subjectieve en objectieve aspecten van sociale en fysieke veiligheid in bos- en natuurgebieden*. Alterra Rapport 1060. Wageningen: Alterra (in Dutch).

Veer, M.M., Abma, R., & van Duinhoven, G. (2006). Openstelling van bos en natuur. *Vakblad Natuur Bos Landschap* 3(3), 2–8 (in Dutch).

Velarde, M.D., Fry, G., & Tveit, M. (2007). Health effects of viewing landscapes – landscape types in environmental psychology. *Urban Forestry and Urban Greening* 6(4), 199–212.

Vera, F. (2005). De sterfte in de Oostvaardersplassen in een internationaal kader. *Vakblad Natuur Bos Landschap* 2(10), 20–22 (in Dutch).

Verboom, J. (2004). *Teenagers and biodiversity – world's apart? An essay on young people's views on nature and the role it will play in the future*. Wageningen: Alterra.

VerHuëll, H.C.A. (Ed.). (1878). *Rapport over het Haagsche Bosch*. In opdracht van Minister van Financiën. 's-Gravenhage (in Dutch).

Vitse, T. (2001). New urban forests in the Desired Flemish Forest Structure – Practical experiences from West Flanders. In: C.C. Konijnendijk, & Flemish Forest Organisation (Eds.), *Communicating and financing urban forests. Proceedings of the 2nd and 3rd IUFRO European Forum on Urban Forestry, Aarhus (May 1999) and Budapest (May 2000)* (pp. 81–85). Brussels: Ministerie van de Vlaamse Gemeenschap, Afdeling Bos en Groen.

von Gadow, K. (2002). Adapting silvicultural management systems to urban forests. *Urban Forestry & Urban Greening* 1(2), 107–113.

Waggoner, P.E., & Ovington, J.D. (1962). *Proceedings of the Lockwood Conference on the Suburban Forest & Ecology*. March 26, 27, 28, 1962, New Haven, Connecticut. The Connecticut Agricultural Experimental Station Bulletin 652. New Haven.

Walden, H. (2002). *Stadt – Wald. Untersuchungen zur Grüngeschichte Hamburgs*. Beiträge zur hamburgischen Geschichte Bd. 1. Hamburg: DOBU-Verlag (in German).

Walker, G.B., & Daniels, S.E. (1997). Foundations of natural resource conflict: conflict theory and public policy. In: B. Solberg, & S. Miina (Eds.), *Conflict management and public participation in land management* (pp. 13–36). EFI Proceedings No. 14. Joensuu: European Forest Institute.

Walmsley, D.J., & Lewis, G.J. (1993). *People & environment – Behavioural approaches in human geography*. Harlow: Longman Scientific and Technically.

Walraven, G. (2006). *Geef burgerschap een sterk stedelijke basis. De stad als opstap naar verbondenheid*. Rotterdam: Hogeschool INHOLLAND (in Dutch).

Walsh, V., & Goodman, J. (2002). From taxol to Taxol®: The changing identities and ownership of an anticancer drug. *Medical Anthropology: Cross Cultural studies in Health and Illness* 21(3–4), 307–336.

WEBPOL. (2006). *OMO undersøgelse om danske børnefamiliers forhold tilsnavs og fysisk aktivitet*. Retrieved August 7, 2007 from Pyyki Web site: http://www.pyykki.fi/dk/snavsergodt/OMOundersogelse.pdf (in Danish).

Werquin, A.-C. (2004). Leisure activities and natural spaces. Additional information from enquiries, nationally and locally (Marseilles). In: A.C. Werquin, B. Duhem, G. Lindholm, B. Oppermann, S. Pauleit, & S. Tjallingii (Eds.), *COST Action C11 Green Structures and Urban Planning – Final report* (pp. 256–258). Brussels: COST C11. Retrieved October 10, 2006 from Map21 Web site: http://www.map21ltd.com/COSTC11-book/pdfs/e-%20Human%20and%20Policy.pdf.

Westoby, J. (1989). Introduction to world forestry, Oxford: Basil Blackwell.

White, R. (1996). “Are you an environmentalist or do you work for a living?” Work and nature. In: W. Cronon (Ed.), *Uncommon ground. Rethinking the human place in nature* (pp. 171–185). New York & London: W.W. Norton & Company.

Widmer, J.-P. (1994). L’espace Rambouilet. *Arborescences* 1994(53), 42 (in French).

Wiegersma, L, & Olsen, I.A. (2004). *NeighbourWoods – Comparative analysis of tree urban woodlands in Denmark and the Netherlands*. Copenhagen: The Royal Veterinary and Agricultural University.

Wiggins, W.E. (1986). *Ancient woodland in the Telford area. A survey of present day woodland occupying ancient sites*. Telford Nature Conservation Project. Telford: Stirchley Grange Environmental Interpretation Center.

Williams, K., & Harvey, D. (2001). Transcendent experiences in forest environments. *Journal of Environmental Psychology* 21, 249–260.

Williams, R. (1973). *The country and the city*. Oxford: Oxford University Press.

Wing, M.G., & Tynon, J. (2006). Crime mapping and spatial analysis in National Forests. *Journal of Forestry* 104 (6), 293–298.

Wolf, K. (2003). Public response to the urban forest in inner-city business districts. *Journal of Arboriculture* 29(3), 117–126.

Wolschke-Bulmahn, J., & Küster, H. (2006). *Die Eilenriede. Hannovers Stadtwald und der Eilenriedebeirat*. Hannover: Landeshaubtstadt Hannover and Leibniz Universität (in German).

Woodland is earmarked for green burial site. (2004, September 9). *Edinburgh Evening News*. Retrieved December 14, 2007 from Edinburgh Evening News Web site: http://edinburghnews.scotsman.com/ViewArticle.aspx?articleid = 2567329.

Woodland Trust. (2006). *Magical woodland celebrations at Belvoir Forest Park*. Media release, October 2, 2006.

Woods for People. (2007). Retrieved August 3, 2007 from Woods for People Web site: http://www.woodsforpeople.info/.

World Health Organization. (1946). *Constitution of the World Health Organization*. Geneva: WHO.

Worpole, K. (2006). Strangely familiar. In: *East London Green Grid: Essays* (pp. 10–11). London: Mayor of London.

Worpole, K. (2007). 'The health of the people is the highest law'. Public health, public policy and green space. In: C.W. Thompson, & P. Travlou (Eds.), *Open space: People space. Engaging with the environment. Conference Proceedings* (pp. 11–21). London & New York: Taylor & Francis.

Worpole, K., & Knox, K. (2007). *The social values of public spaces*. York: Joseph Rowntree Foundation.

Worpole, K., & Rugg, J. (2007). Places of remembrance. *Green Places* 2007(36), 12–13.

Wytzes, L. (2006). Terug naar de duinen. Elsevier 2006 (May 20), 36–37 (in Dutch).

Yokohari, M., Takeuchi, K., Watanabe, T., & Yokota, S. (2000). Beyond greenbelts and zoning: a new planning concept for the environment of Asian mega-cities. *Landscape and Urban Planning* 47(3–4), 159–171.

Zijderveld, A.C. (1983). *Steden zonder stedelijkheid: cultuursociologische verkenning van een beleidsprobleem*. Deventer: Van Loghum Slaterus (in Dutch).

Zollverein. (2006). *Zollverein 31/8*, Das Magazine, Ausgabe 2006, pp. 30–31 (in German).

Zollverein. (2007). Retrieved August 6, 2007 from Zollverein Web site: http://www.zollverein.de/ (in German).

Zuid-Hollands Landschap. (2006). Klimaatbosjes. *Zuid-Hollands Landschap* 2006(4), 24 (in Dutch).

Zürcher, E. (2004). Lunar rhythms in forestry traditions – Lunar-correlated phenomena in tree biology and wood properties. *Earth, Moon and Planets* 85–86, 463–478.

Zürich will Naturpark – so oder so. (2006, July 18). *Zürichsee-Zeitung* (in German).

Index

A

Aalto, Alvar 9
Aarhus 104
access 16, 39, 44, 73, 78–79, 84, 131, 133, 139–143, 151, 155, 162, 165–168, 175, 177, 184, 192, 199–201, 204
  physical – 139–140, 168, 175, 199–201, 206
  public – 73, 84, 140, 151, 162–163, 167–168, 175, 184, 206
  social – 79, 140, 167–168, 175, 177, 199–201, 206
Act of Redemption (Acte van Redemptie) 68
Adirondacks 85
adult 44, 91, 127, 137, 201
Aesculus hippocastanum L. 144
aesthetics 3, 6–8, 60, 100, 102, 204
  forest – 100
afforestation 62, 76, 115, 153, 157
agriculture 3, 54–57, 65, 158, 185
Ailanthus altissima (Mill.)Swingle 146
air pollution 127, 132
Alkmaar 57
Alkmaarderhout, Alkmaar 45–46, 57, 101, 115–116
allergy 43
allotment garden 112, 148, 164
Almere 108, 179
Alnus spp. 27
Alphand, Adolphe 67
Amazonia 77, 121
Amelisweerd, Utrecht 179–180
Amsterdam 1, 7, 9, 59, 82, 92, 137, 164, 169, 179
Amsterdamse Bos, Amsterdam 1, 14, 59, 69, 88, 92, 101, 114–115, 148–149, 164–166
amusement park 14, 23, 33, 75, 87, 101, 166, 170
animals 15, 31, 35–36, 42–45, 50, 65, 81, 86, 111–112, 117, 135, 182
  wild – 112, 135
Anspach, Jules 74
Antarctica 121
Anuradhapura 22
Apeldoorn 68, 167
Aquila chrysaetos L. 186
arboretum 23, 86, 144–146, 150
archetype 19
Argentina 155
aristocracy 6, 16, 49, 51–53, 64–66, 71–74, 83, 100, 112, 144, 147–148, 162, 169, 172
Armenia 60
Arnhem 54, 74–75, 99, 104, 106, 151, 154, 167, 178, 183
Arnos Vale Cemetery, Bristol 24, 47, 79, 171
art 2, 4, 9, 16, 82, 87, 97–109, 113, 129, 147, 173, 201
  land – 103, 108
ash 27
Asia 144
Athens 8, 54, 59
Atlantic cedar 100
attack 38, 68, 119
attention restoration 136
Australia 43–44, 117, 155
Austria 28–29, 127, 154–155
Ayazmo Park, Stara Zagora 119, 171
Ayer Hitam Forest Reserve, Kuala Lumpur 62, 132

B

Baarnse Bos, Baarn 147
Baden-Meinhof group 39
Baden-Wuerttemberg 51
Bahçeköy Forest, Istanbul 146

Bahrdt, Hans Paul 169, 191
Bakken, near Copenhagen 23, 166, 170
bald cypress 144
Balkan 60
Bambësch forest, Luxembourg 89
Barcelona 86
Basel 55, 61–62
behaviour 1, 4, 6, 12, 16, 35–36, 38, 45–47, 63, 73, 83, 87–88, 90, 95, 136, 158, 169, 175, 186, 193, 203
  anti-social – 4, 46, 73
Beirut 60
Belém 77
Belfast 8, 32–33, 53, 99, 102–104, 144, 150, 196
Belgium 41, 53, 74, 86, 101, 115, 120, 127, 135, 146–147, 168, 172, 174–175, 179–180, 205
Belvoir Park (Forest), Belfast 32, 53, 99, 102, 144, 147, 150, 172
Bentwoud, Zoetermeer 102–103
Berg en Bos, Apeldoorn 167
Berlin 1, 4, 10, 31, 51, 53, 56, 59, 65, 72, 75, 77, 84, 86, 100, 111, 125–126, 146, 149–150, 153, 163–166, 183
Berliner Forstschule 146
Berliner Tageblatt 72
Berliner Volkszeitung 72
Berne 55
Beuys, Joseph 99–100
Bible 21
biodiversity 102, 112, 115, 123, 194, 196
Birmingham 98
Black County Urban Forest 196
black locust 146
Blom, Holger 130, 148
Board of Conservators of Wimbledon and Putney Commons 62
Boboli Gardens, Florence 66
Bo(dhi) tree 22
Bois de Boulogne, Paris 1, 10, 23, 40, 65, 67, 69, 84, 87, 98, 131, 163
Bois de la Cambre (Terkamerenbos), Brussels 74
Bois de Vincennes, Paris 10, 23, 40, 53, 65, 67, 75, 84, 131–133
Bold Colliery, St. Helens (Merseyside) 76
Boletus edulis Bull.:Fr. 62
Bosnia(-Hercegovina) 59, 172, 174
Bos primigenius Bojanus 144
Bos ter Rijst, Schorisse (Maarkedal) 53
bourgeoisie 6, 53, 67, 72, 84, 100, 113, 162
Brandenburger Kurfürsten 56, 65
Brazil 77
Breda 53, 69–70, 74–75, 83, 94, 181
Bremen 39
Bristol 24, 47, 79, 171
Brno 104, 147
Brøndbyskoven, Copenhagen 15, 90
Brown, Lancelot ('Capability') 101, 113–114
Bruges 179–180
Brussels 23–24, 52–53, 74, 99, 113, 145–146, 150, 174–175, 183
brownfield 111, 116, 125, 158, 193
  land 193
  sites 111, 116, 125, 158, 193
Buddha 22
Bulderbos, Hoofddorp 179
Bulgaria 119
BUND 125
business 4–5, 14, 75, 133, 179–180, 191
  parks 14, 133
Butchers' Wood, Dublin 40
Byron, George Gordon 98

C

CABERNET 193
California 43, 67
Camerton 202
Camerton Batch Local Nature Reserve, Camerton 202
campaign 135, 140, 158, 172, 180
Canada 45
Canadian Model Forest Network 151
Canberra 117
Castelfusano, Rome 65, 95
Castelporziano, Rome 67
Castlereagh Borough Council 144
Cedar of Lebanon 100, 144
Cedrus atlantica (Endl.)Carrière 100
Cedrus libani A.rich 100, 144
cemetery 24–25, 47, 79, 171
  forest – 24
Central Park, New York 76, 85, 87, 172
Cervus elaphus L. 117
Charles I of England 53
Chernobyl 43
Chicago 47
child 33, 91, 134, 154–156
  's play 91, 134
children 17, 26, 32–33, 43–44, 47, 89, 91–92, 95, 108, 134–135, 137, 143, 153–159
Christchurch 119, 124
Christian V of Denmark 53
Christmas tree 27
Cicero 54

city
authorities 56, 72–75, 114, 116, 164, 186
ephemeral – 5–6, 75
government 117
marketing 75, 77
parks 14, 25, 39, 77–78, 82, 88, 100, 119, 164, 166, 169, 173, 179, 196, 202
post-industrial – 74
regional – 194
City Beautiful Movement 130
city council of Freiburg 57
city council of Zurich 57
city forest
heritage 10, 17, 206
landscape 149, 174, 193
city forestry 10, 17, 32–33, 44, 49, 57–61, 64, 79, 99, 106, 109, 114–115, 120, 146, 150, 157, 161, 171–172, 175, 184, 186, 193, 196, 198, 201, 205–206
history 10, 64
City of London Corporation 58, 72
civilisation 1, 8, 13, 15, 20, 28, 54, 81–82, 94, 111, 122–123, 203
Chigwell Row Wood, London 107
China 76
class (social) 8, 49, 53, 64, 66–68, 72–73, 78, 83–84, 128–129, 162–164, 170
fight 72
clergy 64–66
Cleveland 88
climate change 131, 158
close-to-nature forest management 68, 114, 147
commerce 4–5, 7
commercialisation 191
commitment 17, 129, 181, 186–187, 198–199, 202
common 8, 40, 51–52, 56, 62, 64–65, 71–73, 94, 129, 161–162, 169, 171, 177
lands 50, 72, 161
rights 40, 50, 51, 168
wooded – 8, 50–51, 56, 64–65, 71–73, 161–162
Common Ground 172
Commons Preservation Society 72–73
Commons Protection League 72–73
community 5, 7, 10–14, 17, 31–33, 44–47, 58, 61–62, 76, 79, 106–108, 118–119, 122, 140, 156, 167, 169–171, 173, 175, 179–180, 191, 193, 196–198, 201–202
building 97, 161, 199
forest 58, 61, 99, 106, 108, 149, 165, 195–196, 202, 205
forestry 196, 202
group 47, 79, 194
identity 161, 170–175, 198
involvement 33, 47, 79, 196, 206
sense of – 199, 202
competitiveness 192, 193
concert 77, 87, 166–167, 183
conflict 4, 17, 25, 43, 50, 52, 57, 60, 62, 65, 72, 79, 84, 94, 101, 107, 117–118, 120, 122, 129, 138, 150–151, 157, 171–172, 176, 177–187
social – 4, 129, 177
management 178, 186–187, 205
recreational – 178, 183–185
urban development – 178–179
Connecticut 149–150
connoisseur 151
Constable, John 98
Copenhagen 1, 15, 23, 25, 53, 62, 85, 90, 97, 113, 140, 163, 166, 169, 185
coppice 50, 146, 194
Corylus avellana L. 21
COST 136
Cowpen Bewley 33
Crataegus monogyna Jacq. 27
crime 38–41, 45–47, 79, 130, 155, 201
prevention 38, 44
criminal behaviour 16, 35, 38, 45
crisis 41, 55, 59–60, 65, 191
Croatia 60
culture 1–5, 8–14, 16, 19, 20–21, 25–28, 32, 42, 76, 81, 87, 92–93, 97–100, 104–106, 111, 118–121, 147–148, 170, 172–175, 180, 184, 190, 194, 196, 199, 200, 203–204
city 173
national – 8, 98, 147, 173, 175
forest – 8, 199
cultural
diversity 2, 87
historical values 100, 103, 206
landscape 1, 10–11, 17, 63, 93, 115, 120, 161, 189, 194
x *Cupressocyparis leylandii* 'Robinson Gold' 144, 172
Cupressus spp. 54
customary rights 50, 161
cycling 82, 85–86, 129, 132, 166, 167–168, 175
Czech Republic 46

D

dance 29, 78, 107–108
dancing 104, 167
Dama dama L. 86

Dauerwaldvertrag 163
De Balij-Bieslandse Bos, Zoetermeer 102
defensible space 45
democratisation 8, 64, 67, 71–73, 149, 169, 177
demonstration 17, 100, 143, 150–153
  forest 17, 143, 150–153
Denmark 25, 30, 61–62, 88, 91, 103, 115, 120, 131, 135, 137, 140, 152–153, 155–156, 165, 167, 184
Depression 59
design 6, 8–9, 14, 16, 38, 44, 61, 64, 74, 90, 92, 97, 100–103, 113–114, 120, 126, 140, 147–150, 157–158, 181, 200
Des Moines 35
disease 42–44, 116, 119, 127–128, 131–135
  welfare – 128
Disney World, Orlando 101
Djurgården (also Norra Djurgården), Stockholm 1, 23, 65, 69, 76, 78, 83, 97, 133, 146, 154, 156, 164, 166, 174, 178, 196–197, 200
Djurgården Parliamentary Committee 178
dogs 101, 182, 200
domestication 3, 126
  of nature 126
Donau-Auen, Vienna 179
Douglas, David 144
Douglas fir 144
drinking water 58, 62, 115, 189, 206
Dublin 40–41, 67, 167
Dubrovnik 60
Dusseldorf 100
Dutch East India Company 144
Dutch State Forest Service (Staatsbosbeheer) 90, 92, 108, 117, 151
Dutroux, Marc 41
dwelling 13, 20–21, 90, 122, 178

E

ecological transformation 3
ecosystem 115, 118, 192
Ede 151
Edinburgh 21, 25, 124
Edo (Tokyo) 148
education 17, 44, 140, 143, 147, 151, 153–159, 203
  environmental – 17, 155–159
Eelde 42
Efteling, Kaatsheuvel 90, 101
Eilenriede, Hanover 28, 39, 42, 51, 57, 69, 74, 83, 92, 100
elderly 29, 42, 121, 133–135, 199
Elict, Thomas Lamb 178
elite 3–4, 7–8, 52–53, 64, 67–69, 71–72, 78, 83–84, 87, 120–121, 129, 143, 162–164, 189
  city forests 67
Emscher Landscape Park, Ruhr area 153, 195–196, 203
England 33, 41, 52, 54–55, 61–62, 64–65, 71, 76, 91, 93, 99, 100–101, 105, 131, 148, 155, 168, 171, 173, 200
English Community Forests 149, 195–196, 202
Enlightenment 3, 7, 113, 118
entertainment 5, 48, 75, 90, 95, 166, 170, 191
  landscape 90, 101
environmental 4, 11–14, 17, 25, 44, 60–61, 64, 70, 74, 76–79, 108, 111, 114, 118, 125, 127–128, 130, 132, 136, 138, 141, 143, 147, 155–159, 168, 171, 179, 181–182, 189, 193–197, 200, 202, 205
  justice 64, 78–79
  NGO 108, 125, 179
  psychology 12, 130, 136
  services 60–61, 132, 141, 193, 205
Epping Forest, London 1, 23, 38, 40–41, 50–52, 58, 69, 71–72, 84, 94, 102, 108, 122, 149, 163, 166, 168, 179
Epping Forest Act 108
equity 120, 192, 199
Erfurt 57
Ericsson 73
escape 8, 16, 36–38, 41, 47, 58, 66, 81–95, 124, 129, 140, 162, 164–165, 169–170, 193–194, 203
Essen 104
estate 8, 22, 47, 53–54, 65–67, 72–73, 78, 84, 102, 107, 143–144, 147–148, 184
Estonia 23
ethnic 11, 42–44, 78, 87, 92–93, 155, 170, 199, 201
  groups 11, 42
  minorities 43, 78, 93, 199, 201
Eurasian horse chestnut 144
Europe 2, 6–8, 10–11, 14, 21–22, 26, 28, 33, 42–44, 51–52, 55–58, 62–63, 66–70, 73–74, 78, 85–88, 98–99, 102, 112, 114, 129–130, 138, 140, 144, 146–149, 157, 161, 163–166, 171, 180, 182, 186, 192–193, 199–200, 205
European larch 146
European Union (EU) 127, 179, 192
Evelyn, John 9, 21, 55
excitement 16, 35, 44–48, 203, 205

exotic 24, 65, 86, 111, 113, 116, 119–120, 143–146
species 111, 116, 144–146
trees 24, 120, 143–146
vegetation 119

F

fair 87, 166
fairy tale 28, 31, 33, 36, 41, 43–44, 90, 104
Faith Wood, Tees agglomeration 33
fallow deer 86
fashion 6–8, 16, 33, 97, 100–101, 147–149, 201
fear 5, 16, 29–31, 35–48, 71, 79, 86, 93–94, 107, 112, 116, 119, 121, 125, 135, 158, 175, 201, 203
primeval – 16, 35–38
of nature 42–43
festival 33, 77, 83, 87, 99, 104, 107, 162, 175
Ficus religiosa L. 22
Filbornaskogen, Helsingborg 158–159
Findeisen-Fabeyer 39–40
Finland 9, 78, 88, 91, 122, 164, 174, 178, 196
fire 3, 36, 43–44, 55, 57, 60, 116, 119, 205
Fischer, Jonathan 99
fishing 61, 83
Flanders 53, 92, 115, 120, 135, 147, 205
Flemish Forest Organisation 92, 108
Flemish Minister of Environment 180
Flevohout, Lelystad 151
Flevoland 102–103
Florida 39, 138, 190
folklore 26, 28, 32, 36, 41, 173
Fontainebleau, near Paris 62, 67, 83–84, 91, 99, 102, 113, 149, 167, 181
Forêt de Saint-Germain, Paris 53, 61, 75, 149
Forêt de Sénart, Paris 155
forest
biomass 61
cemetery 24
community – 58, 61, 99, 106, 108, 149, 165, 195–196, 202, 205
edge 19, 36
feeling 15, 88
management 45, 48, 56–57, 61, 67–68, 74, 77, 114–115, 126, 146–153, 158, 178, 181–183, 185, 187, 201
ownership 56–58, 202
planning 33, 102, 186
products 16, 49, 60–62, 132, 162, 185
school 44, 146, 155, 157
services 49, 58
social – 161–176
users 15, 36, 42–43, 65, 82, 90–91, 121
Forest of Dean, Gloucestershire (England) 39
forester 17, 23, 41, 44, 61, 68–69, 104–105, 107, 120, 125, 146, 149–151, 155–158, 176, 181, 194
Forest Faculty, University of Istanbul 146
Forest of Belfast, Belfast 104, 196
forestry
city – 10, 17, 32–33, 44, 49, 57–58, 60–61, 64, 79, 99, 106, 109, 114–115, 120, 146, 150, 157, 161, 171, 172, 175, 184, 186, 193, 196, 198, 201, 205–206
community – 196, 202
profession 146, 181
Forest Service of Oslo 181
Forest Stewardship Council 61, 78, 147
Forn Siðr 31–32
Forstenrieder Park, Munich 29, 65, 112, 147
Foster, Norman 104
France 6, 10, 28, 53, 57, 65, 67, 69–70, 84, 88, 90–92, 99–100, 113, 129, 149, 163, 173, 181
Frankfurt 70, 201
Frederick II of Prussia 84
Frederick the Great (also Frederick II) of Prussia 53
Frederick William I ('The Soldier King') of Prussia 53, 65
Freiburg (i. Br.) 10, 51, 57
French State Forest Service (Office National des Forêts) 61, 149, 155, 181
Friedrich, Froebel 154
frontier 11, 47, 82, 94, 115, 193
Forest Park, Portland 56, 114, 179
Framtidens Skog, Umeå 151
fringe 8, 15, 36, 38, 82–83, 94, 115, 122, 193, 203
urban – 8, 82–83, 94, 115, 193
Fugleberg 186

G

game (animals) 33, 52–55, 60, 62, 65–68, 86, 112, 147
keeper 65
park 86, 147
garden 29, 31, 49, 53, 60, 66, 71, 82–84, 86–88, 90, 98, 100–101, 118, 120, 127–129, 134, 138, 144, 148, 152, 158, 162–164, 166–167, 204
allotment – 112, 148, 164
baroque – 100–101
healing – 134
zoological – 84, 86–87, 90, 101, 158, 166–167, 204

Garden Cities 130, 148
Gavnø 184
Gdansk 49, 52–53, 55, 84
gender 170
Genesis (British pop band) 41
Germany 10, 21, 31, 36, 43, 45, 51, 53, 56–60, 62, 65, 68, 70, 83, 89, 97, 99–101, 111–112, 119–120, 129, 134, 146–148, 173, 182–184, 186, 195, 203
Gilgamesh 27
Ginkgo 144
Ginkgo biloba L. 144
Glasgow 21
Glastonbury Thorn 27
globalisation 5, 17, 74, 189, 198
golden eagle 186
golden rain tree 144
Goering, Hermann 53, 68
Goettingen 10, 151
Grafenberger Wald, Düsseldorf 100
grazing 49–52, 57, 161–162
Great Britain (also: Britain) 4, 6, 9, 25, 41–43, 47, 55, 61, 71–72, 98, 100, 108, 111, 114–115, 120, 128–130, 144, 154, 162, 175, 191, 196, 202
Great Fire of London 55
Greece (Ancient) 21, 54
green belt 104, 151, 174
Greenwood 26, 98, 173
Grimm Brothers 28, 36
Grönna Lund, Stockholm 166
Groundwork St. Helens 76
grove 16, 19, 21, 23, 49, 54, 78, 103
Grunewald, Berlin 1, 51, 65, 72, 75, 84, 149, 154, 163
Gustavsson, Roland 153

H

Haagse Bos, The Hague 38, 68–69, 72, 87, 98, 166, 177
Haarlem 41, 57, 68, 83, 98, 147
Haarlemmerhout, Haarlem 41, 57, 68, 83, 98, 101, 147
habitat 43, 52, 107, 111, 116, 158
Hamburg 10, 32, 39, 58, 148, 165
Hampstead Heath, London 38, 40, 72, 75, 83–84, 98, 104, 114, 163
Hampstead Heath Protection Society 114
Hanebuth, Jaspar 39
Hanover 10, 28, 39, 42, 51, 57, 69, 74, 83, 92, 100
Hartig, Georg Ludwig 146
Hastings 55
Hausmann, Georges-Eugène 148
hawthorn 27
hazel 21
health 7, 17, 26, 29, 31, 43, 62, 71, 73, 76–77, 79, 86, 107, 116, 127–141, 148, 155, 163, 165, 174, 191, 198, 200–203, 206
  care 133–134, 138
  mental – 128, 133–139, 174
  physical – 17, 128, 130–135, 138
  public – 127–131, 135, 138, 140, 163, 200, 206
Heck cattle 144
Hellasreservat, Stockholm 164
Helsinki 73, 77, 86, 88, 91–92, 100, 113, 122, 137, 164–166, 168, 170, 174, 178, 182
Henry I of England 50
Henry III (Count) of Holland 83
Henry VIII of England 52
heritage 10, 17, 22, 24, 31, 36, 71, 98, 104, 108, 173, 195, 206
  national – 173
  cultural – 31, 98
  industrial – 108, 195
Herne (the Hunter) 28
's-Hertogenbosch (Den Bosch) 78
Highgate Wood, London 46, 65, 114
Highland cattle 117
Hildebrand (Nicolaas Beets) 98
hiking 83, 89, 91, 129
Hiroshima 70
Holland 4, 56, 98
holly 49
Holstebro 30, 153
Holzfrau 28
Hood, Robin 25, 40, 94
Horsh Beirut Forest, Beirut 60
Horsterwold, Zeewolde 102
Horticultural Society 144
hospital 23, 29, 128, 133–134
Howard, Ebenezer 148
Hunt, Leigh 98
hunting 3, 8–9, 14, 16, 20, 28, 38, 40, 49–53, 60–62, 64–68, 72, 74, 78, 83–87, 100–101, 112, 120, 128, 146–148, 162–163
  domain 8, 28, 52–53, 64–66, 84, 86–87, 100, 120, 147, 163
  reserve 14–84
Hyde Park, London 95, 98, 162

I
Identity 1, 5–8, 11–13, 17, 32, 76–77, 106, 119, 122, 155, 157, 161–162, 170–177, 186–193, 198, 200, 206
community – 161–162, 170, 172–175, 198, 202
local – 1, 7, 17, 161, 174–175, 189, 191–192, 198
national – 6, 11, 173–174
Ilex aquifolium L. 49
illness 127, 136
Industrialisation 7, 53, 56, 58, 84, 112–113, 128, 162
industry 51–52, 54–56, 104, 108, 128, 133, 195, 200
infrastructure 7, 14, 44, 83, 140, 151, 165–166, 179–180, 184, 189, 193, 196–197, 206
green – 189, 196–197, 206
recreational – 14, 83, 151, 166
involvement
community – 33, 47, 79, 196, 200
public – 74, 171, 176
Iowa 35
Ireland 26, 108, 115, 120, 185, 202, 205
Iskælderskoven, Fugleberg 185
Israel 41, 70–71, 87, 132, 156
Istanbul 146
IUFRO 130

J
Jægersborg Dyrehave, Copenhagen 1, 23, 53, 85, 97–98, 113, 146, 163, 165–166, 169–170
Jaguar (car company) 61
Jaguar Lount Wood, Lount 61
Jerusalem Peace Forest, Jerusalem 41, 70
Joenkoeping, Sweden 126
Joseph of Arimathea 27
Juglans regia L. 158
Juglans spp. 61, 158
jungle 122
urban – 122

K
Kalevala 28
Kapucijnenbos, Brussels 23
Kathmandu 117
Katyn Forest, near Smolensk 41
Keats, John 98
Kennington Common, London 71
Kensington Gardens, London 53
Keskuspuisto, Helsinki 166, 182
Kiefer, Anselm 99
Kings Wood, London 39
Klosterwald, Tallinn 23
Knossos 8, 54
Koelreuteria paniculata Laxm. 144
Konstanz 22–23, 45, 89, 134
Koolhaas, Rem 104
Kos 21
Kralingse Bos, Rotterdam 101, 148, 164
Kuala Lumpur 25, 62, 131

L
Lainzer Tiergarten, Vienna 154
land art 103–104, 108
landscape
architecture 130, 206
cultural – 10–11, 63, 120, 161, 189, 194
design 90, 100, 113, 149
entertainment – 90, 101
forest – 3, 11, 13, 16, 37, 61, 149, 174, 193
industrial – 195
laboratory 17, 152, 153, 157, 205
planning 153, 205–206
post-romantic – 101
therapeutic – 17, 128, 136–137, 206
urban forest – 195–196, 199, 205
style 100–101, 113–114, 148
Lappersfortbos, Brugge 179–180, 186
Larix decidua Mill 146
Las Kabaty, Warsaw 178
learning 17, 44, 82, 92, 114, 124, 143–159
forest 92, 114
Lebanon 60
legislation 102, 124, 196
legitimacy 181
leisure 6, 36, 48, 58, 66–67, 71, 73, 76, 81–88, 108, 115, 129–130, 161–162, 164–166, 168, 170, 174, 181, 184, 200–201, 203, 205
Lelystad 151
Leuven 86
lifestyle 2, 47, 61, 65, 88, 90, 94, 127, 193, 201
lime (tree) 57, 108
Lincoln Park, Chicago 158
Lindulovskaja Roshcha, St. Petersburg 178
Little Red Riding Hood 36, 41
local
distinctiveness 29, 47, 172
knowledge 151
Lombardy poplar 71, 103

London 1, 9, 23, 38, 40, 42, 46, 50, 55, 61–62, 65, 69, 71–73, 78, 83–84, 94–95, 98, 102–104, 114, 125, 129, 144, 149, 162–163, 168–169, 174, 177, 179, 197
London Greenbelt 73
London Natural History Society 114
London plane (tree) 144
London's Green Grid 197
Lorettowald, Konstanz 22, 45, 89
Loriana 22, 45, 89
Losiny Ostrov, Moscow 115, 125
lost (getting) 35–37, 47, 122
Louis-Napoléon Bonaparte of France 67
Lount 61
Louv, Richard 91, 134, 155
Luebeck 151
Luxembourg (city) 89
Lynch, Kevin 119, 155

M

Machiavelli 71
MacKaye, Benton 194
Magistrate of Berlin 72
Malaysia 25, 62, 132
Mamre 21
manifestation 63
Marabou 73
marginalisation 38, 79
market 4, 71, 169
Marseilles 88, 92
Masaryk Forest, Brno 104, 147
Mastbos, Breda 69, 74, 94, 181
masting 51, 57
Maximillian II, emperor of the Austrian empire 28
McArthur, Malcolm 40
media (news) 38, 43, 68, 104, 153, 180, 182
Mediterranean 43
meeting place 75, 95, 148, 161, 169
Menzies, Archibald 144
Merseyside 76
Mesopotamia 54
Metrobosco, Milan 193, 196
Mexico 155
Michaux, Stéphane 41
Middle Ages 3, 6, 23, 49–50, 53, 57, 65, 68, 83, 147
military 64, 67–71
Millington, Thomas 36
mining 53, 57, 76, 153
Ministerial Conference for the Protection of Forests in Europe (MCPFE) 114
minorities (ethnic) 43, 78, 93, 199, 201
mobility 84–86, 165, 175
monastery 6, 21, 23
monument 102–104, 199
Moore, Henry 104
mortality 118, 133
Moscow 66, 68, 73, 78, 86, 115, 125
muflon 86–87
multiple use 114, 147, 182, 205
Münchener Stadtwald, Munich 58
Munich 29, 58–59, 65, 111–112, 147
murder 39–41, 172
museum 5, 75, 87, 90, 104, 149, 154, 158, 180, 204
Museumbos, Almere 103, 108
myth 3, 16, 19, 25–29, 33, 41, 43–44, 97, 121, 173
mythology 26–28, 36

N

Napoleon I of France 65
National City Park, Stockholm 78, 196–197
national
  culture 8, 98, 147, 173, 175
  identity 6, 11, 173–173
  park 67, 78, 86, 111, 115, 120–121, 124–125, 173, 179
  urban park 196, 205
National Health Service (UK) 127, 133
National Institute of Public Health (Denmark) 135
nation building 164
native 26, 113, 124, 131, 136, 138–140, 156, 158
  trees 115
  vegetation 119
nature
  conservation 8, 78, 112–116, 118, 149, 178–179, 181–182, 185–186
  development 115–116, 203
  nearby – 8, 83, 113, 124, 131, 136, 138–140, 156, 158
  observation 86
  'otherness' of – 26, 82, 119, 123
  reserve 107, 111, 113, 117, 202
  's dangers 42–43
  view 100, 107, 111–112, 118, 120–121, 161, 182, 203–204
neighbourhood 5, 31, 73, 78, 82–84, 86, 89, 95, 118, 122, 135, 164, 171–172, 202, 206
NeighbourWoods 158, 171
Neri, Giancarlo 104

Netherlands (The) 6, 14, 25, 33, 38, 41, 43, 57, 69, 74–76, 83, 88, 93, 99, 101–104, 106, 108, 115–120, 125, 131, 133, 136, 144, 147, 151, 154, 158, 164–166, 167–168, 175, 179–181, 184, 200, 203
Newbury 180
Newbury Forest, Newbury 179
new heathens 31
New Town 148, 164, 175
New York 76, 82, 85, 87, 119, 172
New Zealand 32, 119, 124
Nijhoff, Isaac Anne 98
Nordic (countries) 8, 27, 31, 44, 86, 155, 168
North America 43–44, 46, 62, 76, 78, 121, 136, 138, 144, 148, 154, 173
Northern Ireland 8, 32–33, 53, 99, 104, 147, 150, 196
Norway 62, 182, 186

O

oak 21, 51, 60, 68, 71, 100, 170, 181
Odenwald, near Mannheim and Frankfurt 21
Oehlenschläger, Adam 98
Ogham Alphabet 26
Olmsted, Frederick Law 9, 87, 113–114, 120
Olmsted, Frederick Law Jr. 113–114
Olmsted Woods, Washington D.C. 23
Oostvaardersplassen, Lelystad 117–118
Oregon 9, 39, 114, 178
Orlando 190–191
Ortigo, Ramalho 98
Oslo 85, 168
Oslomarka, Oslo 85, 181–182
overcutting 58, 60, 69, 113
overexploitation 8, 52, 54, 60, 69, 107
outlaw 40, 94, 173
ownership 1, 8, 14, 51, 56–58, 64, 69–70, 72, 84, 106, 138, 140, 161, 186, 202, 205
  municipal (forest) – 8, 14, 51, 56–58
  private – 84

P

painter 16, 97–100, 113
painting 97–100
Paradise 21, 25, 47, 121
Parco Nord, Milan 193, 195–196, 205
Paris 1, 10, 40, 53, 61, 65, 67, 69, 75, 84, 90–91, 99, 113, 127, 129, 133, 144, 148–149, 155, 163, 167
park
  city – 14, 25, 38–40, 75, 77–78, 82, 88, 100, 119, 164, 166, 169, 173, 179, 196–197, 202, 204
  municipal – 14, 163, 178
  national – 67, 78, 86, 111, 115, 120–121, 124–125, 173, 179
  royal – 53, 163
  system (movement) 148
Parks and Open Spaces Committee (London) 72
Parque Amázonia, Belém 77
peasants 40, 49–52, 64–65, 78, 94, 120, 161
performance 104, 106–107
Pfeil, Friedrich Wilhelm Leopold 146
Phoenix Park, Dublin 40–41, 67, 167
physical activity 127, 133, 137, 140
picnic 86, 88, 91–92, 165–166
place 1, 3, 5–21, 25, 30–32, 35, 81, 94–95, 106–109, 111, 119, 122, 137, 140–141, 155, 161, 169–178, 186–187, 189–192, 198–206
  attachment 12, 122, 155
  favourite – 137, 175
  identity 11–13, 17, 31–32, 106, 122, 186–187
  making 13, 16–17, 97, 108–109, 119, 161, 169–177, 190, 198–202
  sacredness of – 5, 21
  sense of – 11–13, 31, 198, 202, 206
planning 1, 9, 33, 60, 72–73, 79, 102, 106, 114, 148, 151, 155, 157, 161, 171–172, 176–177, 185–187, 189, 194, 196–200, 205–206
  forest – 33, 102, 186
  landscape – 153, 205–206
  urban – 73, 114, 171
Platanus orientalis L. 144
Pliny 55
pollarding 51
poem 97, 104, 113, 121, 180
poetry 106–107
Poland 52, 84, 155
population growth 3, 21, 52, 55
Populus nigra 'Italica' 71, 103
Porcini mushrooms 62
Portland 56, 114, 179
power 1, 3, 5–8, 16, 21, 27, 31, 49, 52, 55, 61, 63–79, 84, 86, 112, 118, 129, 161–162, 164, 168, 172–173, 177, 184, 186, 190–191, 199
powerscape 63, 68
Prague 52–53
Prater, Vienna 28, 83

predator 42–43, 186
preference 2, 5, 8, 81, 90, 93, 97, 100, 118, 123–125, 151
prestige 8, 53, 64–67, 74, 97, 100–101, 174, 206
Pritzker Family Children's Zoo, Chicago 158
privacy 31–32, 88–89, 184, 203
  reserve 88
privatisation 72, 191, 204
protestors 71, 180–181
Prussia 56, 65, 72, 146
Pseudotsuga menziesii (Mirb.)Franko 144
public
  access 73, 143, 151, 162, 167, 175, 184
  involvement 74, 171, 176
  realm 47, 106
  space 6, 44, 47, 64, 74–75, 104, 140, 161, 169–170, 191, 199
Public Health Act 128

Q
quality of life 79, 86, 123, 127, 129, 153, 189, 192–193, 198
Quellschutzwälder, Vienna 58
Quercus spp. 21, 181
Quercus rubra L. 181

R
Rambouillet, Paris 67, 86, 90
recreation 3, 6, 14, 25, 38–39, 47–49, 53, 58, 60, 64–72, 77, 81, 83–89, 92–93, 100–101, 102, 108, 114–115, 120, 123, 129–130, 148–149, 161–169, 174, 178, 182–186, 194, 196, 203, 206
  (-al) infrastructure 14, 83, 151, 160
  (-al) use 8, 14, 16, 41, 58, 84–85, 87–88, 91, 125, 133, 153, 161–169, 183, 185
  mass – 38, 165–169
  nature-based – 88
Redditch 88, 123, 168
red deer 117
red oak 181
Reformation 23, 56–67
refuge 5, 15, 19, 36, 72–73, 85, 94–95, 98, 118, 169, 173
regeneration 23–24, 114, 116, 123, 146, 174, 181, 182, 193, 195, 197
Reims 103
religion 4, 16, 19, 21–25, 31, 100, 198
Renaissance 3, 66, 83, 101, 112, 148
Repton, Humphry 9, 101, 113–114
Richard I of England 65
Richard II of England 28
Riga 186
Riis Skov, Aarhus 104
ritual 19, 26, 29, 75, 198
Robinia Foundation 151
Robinia pseudoacacia L. 146
Rodin, Auguste 104
role playing 48, 90–91
Romanticism 28, 100–101, 111, 113, 121, 148
Rome 4, 20, 54–55, 66–67, 95
Ronneby 151
Ronneby Brunn, Ronneby 151
Rosh Ha'ayin city forest, Rosh Ha'ayin 132
Rospuda Valley, near Augustow 179
Rotterdam 101, 148, 164
rowan 21
Ruhr area 32, 61, 103, 104, 108, 116, 125, 153, 195, 203
rural 4, 29, 51, 57, 61, 94, 98, 107, 112–113, 125, 153–154, 182, 194, 203, 205
  forest 51
  landscape 107, 194
  life 4, 29, 163
Russia 24, 60, 66, 68, 73, 104, 115, 151, 162, 166, 173, 178
Rye 55

S
Saarbruecken 57
Saarkohlenwald, Saarbrücken 57
Sachsenwald, Hamburg 39
safety 1, 7, 35–38, 42, 47, 91, 135–137, 156, 171, 175, 191, 200
saga 28, 31, 33
Sarajevo 59, 62, 78, 85, 172, 174
satoyama 194
Scheveningse Bosjes, The Hague 95
Schoenbrunn Garden, Vienna 167
school 5, 33, 44, 46–47, 58, 76, 99–100, 108, 112–113, 117, 139, 146, 155–158, 202, 205
  forest – 44, 146, 155, 157
School of Barbizon 99–100, 113
Schorfheide, Berlin 53
Scotland 21, 38, 91, 99
Sea World 204
security 5, 11, 52, 138, 180, 198, 201
sex 47, 94–95
Shakespeare, William 98
Shelley, Percy Bysshe 98
Sherwood Forest, Nottingham 40, 94
shipbuilding 55–56

Sihlwald, Zurich 51, 57, 87, 124–125, 155, 203–204
Sierra Nevada (California) 43
Skogshögskolan 146
Skogskyrkogården, Stockholm 24
Sletten Landscape Laboratory, Holstebro 152–153
Slovenia 55, 57, 147
Smolensk 41
Snogeholm Landscape Laboratory, Snogeholm 153
social
  cohesion 71, 170–175, 189, 191–192
  conflict 4, 129, 177
  control 38, 79, 84, 91, 156, 163, 170
  exclusion 79
  forest 161–176
  resistance 71
  science 206
  stage 161, 167–170
  values 47, 189
Solleveld, The Hague / Kijkduin 125
song 41, 97–98, 121
Sonsbeek, Arnhem 54, 74, 104, 106, 154, 166–167, 178, 183
Sorbus aucuparia L. 21–22
Sosnovka park, St. Petersburg 24
South Korea 138–139
Soviet Union 73, 173
space 1, 4–6, 10–15, 17, 21, 25, 30, 32, 35, 47, 73–75, 78–79, 81, 91, 94, 104, 106, 122, 124, 154, 172, 189–190, 199, 203–205
  public – 6, 44, 47, 64, 74–75, 104, 140, 161, 169–170, 191, 199
  wild- 124–125
  wild adventure – 79, 91, 124, 154
species 6, 26, 78, 111, 113–116, 119, 143–146, 150, 157–158, 174, 185
  tree – 6, 78, 114–115, 119, 143–144, 146, 150, 157
spiritual 2, 10, 14, 16–17, 19–33, 36, 49, 54, 57–58, 82, 97, 103, 107, 118, 129, 137–138, 172, 194, 203
  values 19, 103
sports 77, 83–85, 90–91, 112, 127, 140, 163–164, 166, 174
sprawl 43, 189–191, 193, 205
Sri Lanka 22
stakeholder 64, 171, 192, 205–206
Stara Zagora 119, 171
Stärback, Karl 178
State Forest Technical Academy of St. Petersburg 146
St. Helens 76
St. James's Park, London 162
Stockholm 1, 23–24, 69, 76, 78, 84, 97, 112, 114, 130, 146, 148, 154, 156, 164, 166, 178, 181, 196–197, 200
Stockholm Park System 148
storytelling 26, 107
St. Petersburg (Russia) 24, 59, 66, 68, 73, 75, 104, 146, 151, 162, 166, 178
stress 17, 30, 95, 127–128, 134–141, 158
  relief 95, 137
  recovery 135–136
subsistence 19, 28, 49–51, 62, 117, 128, 189, 202
suburbanisation 189–190
Südgelände Nature Park, Berlin 126
Sungai Buloh, Kuala Lumpur 25
sustainable 114, 191–193, 196, 205
  city regions 192–193, 205
  development 192, 196
  forest management (SFM) 114
  urban development 114, 192
  urban landscapes 192
sustainability 17, 28, 73, 79, 114, 196, 205
Sweden 23–24, 27, 37, 43, 73, 146, 151, 153, 156, 158, 164, 174, 186, 197, 200–201
Swedish Association for the Promotion of Outdoor Life 156
Swedish Society for Nature Conservation 181
Swedish University of Agricultural Sciences 151, 153
Switzerland 55, 57, 61, 124, 147, 202
symbolic values 19, 23, 26, 28, 71

T

Tallinn 23, 52
Tauride Gardens, St. Petersburg 166
Taxodium distichum (L.)Rich 144
Taxus baccata L. 33, 132
territoriality 12–13, 45, 47, 186, 201
Tervuren arboretum, Brussels 23, 145–146, 150
testing grounds 17, 143–150, 157, 205
Thames Gateway, London 196
The Gordel 175
The Hague 38, 68, 78, 87, 95, 98, 125
The Mersey Forest, Merseyside 76
therapeutic landscape 17, 128, 136–137, 206
Tiergarten, Berlin 53, 65, 77, 84, 100, 163–164
Tilia sp. 108
timber 3, 11, 16, 20, 50–62, 65, 68, 112, 124, 143–147, 151, 162–163, 181, 185, 189, 193, 202, 206
  production 53, 61, 143, 147, 181, 202

Tolkien, John Ronald Reuel 98
Tortosa 57
tourism 5, 76–77, 88, 125, 183, 194
Toxovo Demonstration Forest, St. Petersburg 151
Tradescant family 144
training 6, 69, 143, 146–147
  forest 147
transcendence 30–31, 82, 138
tree
  ancient – 21, 29, 98, 101, 102
  Christmas – 27
  exotic – 24, 120, 143–146
  native – 115
  of the Cross 27
  ornamental – 114
  sculpture 104
  species 6, 78, 114–115, 119, 143–144, 146, 150, 157
  working – 49
  World – 26–27
tree of heaven 146
Turkey 146
Turpin, Dick 40, 94

U
United Kingdom (UK) 23–25, 29, 37, 47, 61, 76, 78, 88, 121, 123, 127, 133, 135, 137, 147, 195, 197, 200, 202, 205
United Nations Conference on Environment and Development (UNCED) 114
United States (of America, USA) 23, 29, 45, 56, 76, 78, 88, 113–114, 119, 138, 149, 158, 172–173
United States Forest Service 39, 46
university 26, 111, 146–147, 151, 153
Ur 54
urban
  expansion 12
  forest 1, 9, 78–79, 88, 147, 151, 157, 167, 181, 196, 199, 205
  forest landscape 195–196, 199, 205
  forestry 106, 195, 205–206
  fringe 8, 82–83, 94, 115, 193
  green structure 205
  nature 8, 47, 98, 113–114, 136
  planning 73, 114, 171
  sprawl 189, 193
  values 203
  woodland 9, 29, 37–38, 41–42, 101, 104, 137, 150–151, 157, 166, 168, 171, 177, 193
urbanisation 1, 4, 8–9, 17, 43, 58, 125, 149, 164, 177, 189–191, 200, 205
urbanity 191
Uruk 54
Uutela, Helsinki 170

V
Valkenberg (also Valcberg), Breda 53, 70, 75, 83
vandalism 39, 45–47, 79, 156, 182
Van Goyen, J. 99
Van Schermbeek, A.J. 181
Vargamor 28
Vauxhall Gardens, London 162
Venetian Republic 55
Vestskoven, Copenhagen 185
Victorian era 24, 73, 95, 100, 129, 144, 163
Victory Park 73, 172–173
Vienna 21, 28–29, 58, 74, 78, 83, 88, 90, 102, 123, 154, 165, 167, 179, 182
Villa Borghese, Rome 66
Volkspark 102, 115, 129, 148, 163–164
Volkspark, Hamburg 148
Vondelpark, Amsterdam 82, 137
von Langen, Johann Georg 146
von Thünen, Johann Heinrich 57
Vrchlický, Jaroslav 104
Vrelo Bosne, Sarajevo 62, 85, 174
Vrienden van het Mastbos 74
Vrienden van Sonsbeek 74
Vuosaari, Helsinki 178

W
Wales 22, 39, 121, 155
walking 27, 29, 81–82, 85–89, 98, 129, 132–133, 159, 165, 167–168, 175, 182
Walloon region 41
walnut (tree) 158
Waltham Blacks 23, 40
Wandervogel 29, 129
war 24, 29, 41, 53–54, 56–60, 64–65, 68–70, 73, 101–102, 112, 114, 164, 172–174
Warrington New Town 175
Warsaw 78, 178
Warsaw Nature Conservation Bureau 178
Washington (state) 39
Washington DC 23
well-being 17, 127–131, 135–41, 171, 191–192, 198, 200–201, 203
White Rose Forest, Yorkshire 196
Wienerberg, Vienna 123

Wienerwald, Vienna 78, 86, 88, 102, 155, 165
Wiesbaden-Naurod 29
wild
  adventure space 79, 91, 124, 154
  industrial forest 125
  space 124–125
wilderness 3, 8, 13–14, 17, 28, 30–31, 47, 67, 87, 95, 111–112, 121–125, 137–139, 154, 173, 190, 195, 203–204
  experience 30–31, 87, 95, 125, 137–139
wildlife 111, 125
wildwood 3, 15, 50, 101
William of Orange / William the Silent (of Holland) 68
Williams, Watkin 98
Wimbledon and Putney Commons, London 62
Winchelsea 55
window to the world 81, 86–87, 143
Witte Kinderenbos, Mechelen 172
Winterthur 58
wolf 28, 42–43, 52, 54, 112, 186
wood (timber) 2, 8, 16, 20, 50–62, 69, 114, 123, 151, 194, 206
  construction – 8, 69
  fuel 8, 11, 49–51, 54–60, 69, 124, 189
  production 53, 55, 61
Woodland Trust 33, 168, 196
Woods on your Doorstep 196
working
  class 6, 67, 73, 78, 84, 128–129, 163–164, 174
  environment 76, 124, 136, 202
workplace 124, 133, 139
World Tree 26–27
World War II 24, 41, 59–60, 69–70, 73, 101–102, 114, 164, 173–174
worship 19, 21–23, 31
writer 16, 75, 91, 97–99, 104, 107, 113

Y

yew 33, 132
Yggdrasil 27
Yorkshire 61, 196
Yosemite National Park, USA 67, 113, 120, 173
youths 17, 26, 38, 41, 47, 79, 89, 91, 93, 143, 153–155, 158, 199, 201, 203

Z

Zeewolde 102
Zoetermeer 102
Zollverein, Essen 104, 108
Zoniënwoud (Forêt de Soignes), Brussels 51, 57, 77, 87, 111, 124, 147, 155, 204
zoological garden 84, 86–87, 90, 101, 158, 166–167, 204
Zurich 51, 57, 77, 87, 111, 124, 147, 155, 20